Anonymous

On the Pollution of the Rivers of the Kingdom

The enormous magnitude of the evil, and the urgent necessity in the

interest of the public health and the fisheries for its suppression by

immediate legislative enactment

Anonymous

On the Pollution of the Rivers of the Kingdom
*The enormous magnitude of the evil, and the urgent necessity in the interest of the
public health and the fisheries for its suppression by immediate legislative
enactment*

ISBN/EAN: 9783337237288

Printed in Europe, USA, Canada, Australia, Japan

Cover: Foto ©berggeist007 / pixelio.de

More available books at **www.hansebooks.com**

ON THE

POLLUTION OF THE RIVERS

OF THE

KINGDOM;

THE ENORMOUS

MAGNITUDE OF THE EVIL,

AND THE URGENT NECESSITY

IN THE INTEREST OF

THE PUBLIC HEALTH & THE FISHERIES

FOR ITS

SUPPRESSION

BY IMMEDIATE

LEGISLATIVE ENACTMENT.

*Evidenced by Extracts from the Reports of successive Royal Commissions,
Committees of both Houses of Parliament, Inspectors of Salmon
Fisheries, Medical Officers of the Privy-Council, Registrar General,
&c. &c., presented or returned to Parliament between 1855 and 1868.*

———

CIRCULATED BY THE COUNCIL OF THE
FISHERIES PRESERVATION ASSOCIATION,
23, LOWER SEYMOUR STREET, PORTMAN SQUARE, W.
President—HIS GRACE THE DUKE OF NORTHUMBERLAND.
Vice-President—LORD DE BLAQUIERE.

———

1868.

INTRODUCTION.

THE Council of the Fisheries Preservation Association, in bringing under public notice the subject of the pollution of rivers, deem it unnecessary to use many words of their own in order to secure due attention to an evil which in its wide-spread extent and baneful effects has become one of the deepest *national* importance.

The facts adduced in these pages, taken from the most authentic sources, sufficiently prove its colossal and necessarily ever increasing proportions, and trumpet-tongued proclaim the necessity of some prompt and comprehensive remedial measure, to protect from further injury and destruction the health and lives of the people, and save from further annihilation what, but for these pollutions and other grievous injuries to the river fisheries, would form a very valuable addition to their food.

The Council therefore in the few observations they propose to make, need do little more than point out (but to the important fact they invite special attention), that while the Royal Commissioners and the other authorities quoted have all in the strongest terms denounced the pollution of our rivers by sewage, and mine, and manufacturing refuse, as a most intolerable and dangerous nuisance that must be abated, they one and all at the same time concur in declaring that it can be abated and in a manner satisfactory to all parties; that sewage can easily and profitably, and without danger to the public health be got rid of by application to the land, and that the noxious

A 2

refuse of mines and manufactures can without any serious
interference with the industrial pursuits of the country,
within reasonable limits of expenditure, and even in many
cases with actual profit to the mine owner or manufac-
turer, be disposed of in other ways than by sending it
into the rivers, and thereby poisoning with it, the public,
the fish, the air, and the running waters of the kingdom.

In an Appendix will be found a short statement of the
efforts, commencing in 1855, which have been made to
free our rivers from their dreadful state of pollution.
Though those efforts have, it will be seen, been strenu-
ous and continuous, the Council regret to state that with
the single exception of the main drainage of the metropolis
nothing, absolutely nothing, has yet been accomplished in
the shape of *effective practical legislation* towards putting
down this gigantic and dangerous nuisance, consequently
that nuisance now overspreads the land in all directions,
it being a lamentable truth that (with the one exception
just noted of the Thames at London) there is scarcely a
river, a rivulet, or a brook, contiguous to a population,
or to a manufactory, or a mine, that is free from its pernicious influence.

From the remarks addressed in August last by the
Home Secretary to the deputation which waited on the
Right Honorable gentleman upon this subject from the
Fisheries' Preservation Association, namely, that "*he
"did not intend to continue the investigations, as he be-
"lieved that the experience gained by the inquiries into a
"few rivers would govern the whole,*" the Council were
led confidently to hope that Government would be prepared to introduce this Session a measure adequate to
meet the evil.

In that expectation they have been grievously disap-
pointed, for on the 24th Feb. last, Mr. Hardy informed

the member for Sunderland Mr. Candlish, that " *he was* " *not prepared to legislate on the subject this Session,*" (see Appendix, page 52) and the Home Secretary followed up that declaration by appointing during last month a fresh Commission to continue the inquiries which in the preceding August he then considered had gone far enough, so that it seems but too manifest, that as far as the Government is concerned all legislation in the matter is indefinitely postponed.

Be that however as it may, and be the action or inaction of the Government what it may, the Council on their part will continue their best and most energetic efforts in the cause, until a law has been obtained potent enough to grapple with and put an end to this monster evil, but in doing battle for an object which so vitally concerns the health and interests of the public, the Council feel themselves entitled to look for *the active aid and co-operation of the public,* without which they fear they can do but little, and they would here impress it on all Towns and Constituencies suffering from these pollutions that besides holding public meetings on the subject, in no way can that aid and co-operation be so effectively rendered as by their *petitioning the Legislature for relief and instructing their representatives in Parliament to support such petitions by every means in their power.*

In order to make the fearful state of the rivers of the country more generally known, without the necessity (to acquire that knowledge) of toiling through the voluminous Blue Books, &c., the Council have extracted from those unimpeachable testimonies all the necessary facts of the case, and have embodied them in the compendious form of the pamphlet which they now issue.

This pamphlet is circulated by them in the earnest hope that it may help to evoke such a powerful and decisive

expression of public opinion on the question, that the
Government and the Legislature will be forced to provide,
without more vexatious and needless delay, some thoroughly
efficient remedy for a state of things which, as was most
truly said by the mover* of the second reading of the
" River Waters Protection Bill† of 1865," " *is a disgrace*
" *to a nation that prided itself on its civilization and ad-*
" *vancement.*"

* Lord Robert Montagu. [Hansard, 3rd series, vol. 177, pages
1312—1313.]
† This Bill was withdrawn, but on grounds wholly irrespective of
the magnitude of the evil and the pressing necessity for some legislation to correct it, all which indeed was admitted. [See Appendix,
page 49.]

FISHERIES PRESERVATION ASSOCIATION,
23, Lower Seymour Street,
Portman Square, W.

May, 1868.

DE BLAQUIERE,
President.

The Council have the high gratification of stating that
His Grace the Duke of Northumberland has consented to
preside over their future deliberations,—Lord de Blaquiere
having with the utmost consideration and kindness agreed
to act as Vice-President.

POLLUTION OF RIVERS.

EXTRACTS from the REPORTS of Royal Commissions,
Parliamentary Committees, Inspectors of Salmon
Fisheries, Medical Officers of the Privy Council, Re-
gistrar General, &c. &c., presented or returned to
Parliament between 1855 and 1868.

" TEN years ago," said Lord Robert Montagu, in his
masterly speech when moving, on the 8th March, 1865,
the second reading of the " *River Waters Protection*
" *Bill*," " the Committee on the *Nuisances Removal Bill*
" *of* 1855 had inquired into this subject, and had ascer-
" tained that *our Rivers* had become *absolutely pestilen-*
" *tial*, and were, in fact, *nothing but main sewers*, and had
" urged the Government to take steps for the removal of
" such disastrous influences."
[*Hansard, 3rd Series, vol.* 177, *p.* 1309.]

1855.
Rivers ascer-
tained by
House of
Commons'
Committee of
this year to
have become
pestilential.

In 1858 a Royal Commission on the sewage of towns
reported (page 27) that—
" From the whole of our inquiry we have arrived at the
" following conclusions," one of them being :—

" That the increasing pollution of the rivers and streams
" of the country is an evil of *national importance* which
" *urgently* demands the application of remedial measures ;
" that the discharge of sewage and the noxious refuse of
" factories into them is a *source of nuisance and danger*
" to *health ;* that it acts injuriously not only on the locality
" where it occurs, but on the populations of the districts
" through which the polluted waters flow ; that it poisons
" the water which in many cases forms the *sole supply* of
" the populations for all purposes, *including drinking ;*
" and that it *destroys the fish*."

26 March,
1858. First or
Preliminary
Report of the
Commission-
ers on the
Sewage of
Towns.

4 July, 1860.
Report of
Committee of
the House of
Lords on
Salmon
Fishings,
(Scotland.)

In 1860, among other recommendations, a Committee of the House of Lords recommended (page 12)—

" That penalties to be recovered by *summary process* " be imposed for allowing any refuse matters from any " mill or manufactory to go into any river ;" and

" That a bill be introduced by Her Majesty's Govern- " ment in conformity with the above recommendations."

In 1861 the Commissioners on Salmon Fisheries re- ported (pages 19—21) that—

7 Feb., 1861.
Report of the
Commission
on the
Salmon
Fisheries.
(England
and Wales.)

" The most striking case of contamination of waters by " the efflux from mines was in the *Ystwith* and *Rheidol* " at Aberystwith."

" These two streams both contained salmon in some " abundance thirty years ago. Since the working of the " Goginan lead mines, a *total extinction of animal life* has " taken place in the *Rheidol*. The *Ystwith* has been " *similarly affected* by other lead works." " The most " distinct evidence was given us of the destruction of " salmon from this cause. It was even stated that the " sea-fishery to the extent of some miles out had been " much deteriorated from the same cause."

" No other case of destruction so complete was brought " under our notice ; but there were others of the noxious " effect of *mine waters,* in more or less degree, which, if " continued, must ultimately prove fatal to the fish. " Among these we may mention the *Tawe, Neath, Rhym-* " *ney, Towey, Taff, the South Tyne,* &c."

" In *Cornwall* the salmon fisheries may be said to be " *virtually destroyed* by the mines."

August 1861.
Second
Report of
the Commis-
sioners on
the Sewage
of Towns.

And in the same year (1861) the Commissioners on the sewage of towns in a further report, pages 4 to 9, give a frightful picture, (drawn from their own personal in- spection) of the pollution of the rivers *Irwell, Irk, Roch, Croal, Tonge, Tame,** Mersey, Bollin,* and *Medlock,* saying

---

* In the evidence of Robt. Rawlinson, Esq. (formerly one of the Commissioners on the Sewage of Towns, and also one of the late Commissioners on the Pollution of Rivers), before the House of Com- mons' Committee of 1864, on the sewage question, he says :— " That, before the Tame reaches *Birmingham,* it receives above

of the surpassing filth of the last, where it forms the head of the Bridgewater Canal, " no description can give an " adequate idea ;" adding, that " such is its consistency, " it is said *birds walk over it*," *Derwent, Aire, Calder* and *Don* in their respective courses through or near *Manchester, Middleton, Rochdale, Bolton, Staleybridge, Birmingham, Wolverhampton, Ashton, Stockport, Macclesfield, Walsall, West Bromwich, Bury, Oldham, Derby, Bradford, Leeds, Wakefield, Sheffield, Rotherham, Doncaster,* &c., &c., after which the Commissioners proceed to say :—

" If the discharge of solid matters of sewage and other
" refuse into rivers is prevented, the chief part of the
" inconvenience, which is now rapidly rising to the pro-
" portion of *a national evil*, would at once be arrested.
" *This can be easily and fully accomplished within rea-*
" *sonable limits of expenditure*, and we urge, as the first
" and all important step towards securing this object, and
" the permanent improvement and protection of the rivers
" of the country, that a general local jurisdiction and
" conservancy be created throughout the kingdom, with
" adequate powers and proper guarantees for their due
" administration." Adding (pages 11, 12, 13) that—
" Notwithstanding the incompleteness of our inquiry,
" we trust enough has now been said as to the *enormous*
" *loss* and *injury* produced in various ways, by the pre-
" sent state of neglect and misuse of our rivers, to secure
" *immediate attention* with a view to the adoption of
" *general* and *decided* measures to arrest this *great* and
" *growing evil.*
" As regards the deterioration of water for domestic
" and other useful purposes, the pollution of rivers *is an*
" *evil of immense magnitude.* In extreme cases the water
" is unfit for any kind of use ; but long before this degree
" of foulness is reached, the water has been unfit for
" purposes of cleanliness, and from a very much earlier

" *Birmingham* the sewage of 270,000 people, and all the refuse of
" gas works, pumpings of coal mines, and the drainings of the great
" district of South Staffordshire. *Birmingham* always *suffers* more
" or less from a *type of fever.* Its mortality is much higher than
" that of the Metropolis."

" stage of contamination has been utterly unfit for drinking.
" The last mentioned evil deserves very particular atten-
" tion, *for the danger to health* occasioned by the con-
" sumption of polluted water, is, in our opinion, infi-
" nitely greater than any danger which the effluvia of
" polluted water can occasion. Water, tainted but very
" slightly with sewage, *may determine terrible outbreaks*
" *of disease* among the populations which drink of it* ; and,
" although in such cases as that of Manchester (where the
" water is grossly and offensively foul), there is little
" chance that people will drink that water, yet in cases
" of less obvious contamination, tainted water is, perhaps,
" extensively drunk. Such water, supplied by Water
" Companies, has, in various cases, been suspected, or
" proved to have determined, on a very large scale, the
" distribution of *cholera deaths* during times of epidemic
" visitation ; as in our *South London districts*, during
" the two *last epidemics of cholera ;†* at other times it
" has, probably, exerted equal influence, in determin-
" ing the distribution of deaths from ordinary diarrhœal
" diseases ; and on various occasions it has been shown,
" that what to the common eye is an inappreciable
" pollution of water by sewage, may yet imply very
" serious dangers of infection for the persons who consume
" such water. On these grounds, seeing that brooks and
" rivers are almost universally the sources from which
" Water Companies derive their supplies for large urban
" populations, we deem it to be of *essential importance to*
" *the public health*, that the *running waters* of the country
" should be *strictly protected from pollution.*"

Among other conclusions they arrived at, the Commissioners, at page 39, submitted the following :—

" That this condition of rivers is a *public and national*

---

* As for example : cholera outbreak in 1854 in the Broad Street,
    Golden Square, District.
    Fevers, &c., at Rotherham, 1862-4, [vide Memorial of Rotherham
    Board of Health, page 18.]
    Cholera—East London, 1866.
    Typhoid fever at Guildford, Sept., 1867. [Vide Dr. Buchanan's
    Report to Privy Council, pages 39 and 40.]
    Do. do. at Terling, Dec., Jan., Feb., 1867-8. [Vide Dr. Thorne's
    Report to Privy Council, pages 40—42.]
    Do. do. at Marine Barracks, Stonehouse (Plymouth), Dec. 1867,
    in which case seven marines died—water from well strongly
    suspected. Vide extract from *Lancet*, page 43.
           † Of 1848-9 and 1853-4.

" *nuisance,* it interferes with the convenience and comfort
" of all classes of the people, it damages various and
" important interests as manufacturing establishments,
" canals, *fisheries,* and so on ; it deteriorates property to a
" large extent, and as interfering with a main source of
" water supply is of serious importance to the *public*
" *health.*"

In 1862 the Inspectors of Salmon Fisheries, at page 27,
report that among many other rivers—such as the *Istwith*
and *Rheidol, Pontypool River,* the *South Tyne,* and the
Eden (at Carlisle), &c.—found to be grievously affected
by various forms of pollution,

8th Feb.,
1862.
1st Annual
Report of
the Inspec-
tors of
Salmon
Fisheries.

" The *Calder* is so polluted by dye and print works,
" that the *fish* in it have been *nearly, if not quite de-*
" *destroyed.*"

And in 1863, after referring (pages 59 to 63) to the
way in which the *Kent* by paper works, the *Dovey, Wye,*
Teign, Tamar, Tavey, and *South Tyne* by mines, the
Tawe by copper and other works, and a tributary of the
Usk by chemical works, are severally poisoned and the
fish destroyed, the Inspectors conclude their 2nd Annual
Report by saying :—

2nd March,
1863.
2nd Annual
Report of
Inspectors of
Salmon
Fisheries,
(England
and Wales).

" We are confident that the *injury* done to such rivers
" as are polluted is *capable of great reduction,* and that
" if, by mechanical means, the great proportion of the
" poison from mines or factories can be extracted before
" the fouled water reached the river, the small quantity
" that escapes would be neutralised by the body of pure
" water that receives it. *Such means do exist, and in all*
" cases, in our opinion, *are a source of profit* to their
" employers."

In August, 1863, a deputation from the Sanitary
Associations of Great Britain and the " Fisheries
Preservation Association," waited on the late Lord Palmerston, on the subject of the pollution of rivers and
its prevention, when his Lordship fully recognizing
the extreme gravity of that evil, and the necessity of
putting an end to it, *expressed his intention of bringing*

August, 1863,
Deputation
from the
Sanitary As-
sociations of
Great Britain
and the
Fisheries'
Preservation
Association
to Lord
Palmerston.

forward a Government measure for the purpose,—and had he lived no doubt he would have done so with the same energy his Lordship evinced in suppressing by legislative enactment the smoke nuisance, a nuisance however, which, great as it was, only affected *one* element, while the pollution of rivers, poisons *two*, air as well as water.

His Lordship having requested the deputation to submit its views and wishes in writing, a letter was addressed to him, from which the following are extracts :—

London, 4 March, 1864.

4 March, 1864.
Joint Letter of Lords Ebury and Shaftesbury, on behalf of the Sanitary Associations of Great Britain, and Lords Saltoun and Llanover, the President and Vice-President of the Fisheries Preservation Association, to Viscount Palmerston, First Lord of the Treasury. [Parl. Paper 224, April 20, 1864.]

" ' My Lord,

" When we had the honour of an interview with your " Lordship, you requested us to submit, in writing, the " several propositions we might desire to make, for " preventing the pollution of streams.

" The pollution of rivers and streams has now become " very general, and great injuries result therefrom. These " may be considered—

" First, as affecting the public at large in a *sanitary* " point of view, and what is called in Scotland the " ' amenity ' of the district.

" And secondly, the *fisheries*, many of which have " been totally destroyed by the deleterious matter, which " is thrown into, or suffered to flow into, the rivers and " streams.

" As regards the injury to the *health* and *comfort* " of the population living upon the banks of rivers " and streams, or in their immediate neighbourhood, " it is fortunately unnecessary to use our own language, " because the case has been set forth in its true light, " in clear and unmistakeable terms, in the reports pre- " sented by the eminent men who composed the Sewage " of Towns Commission."

The Letter after quoting from the first report of the Commission on Sewage of towns in 1858, the "conclusion" (before given at length, page 7) come to by those Commissioners, namely, that

" ' The increasing pollution of the rivers and streams " of the country, is an evil of *national importance, which*

" *urgently demands the application of remedial measures,*
" &c.," proceeds.

" As regards injuries to *Fisheries* resulting from the
" pollution of the waters by noxious matters, we have
" only to refer to the very excellent report of the Fishery
" Commissioners of 1860, presented to Parliament in 1861,
" for an exemplification of the manner in which valuable
" Fisheries have been destroyed.

" In that report it is shown that some rivers have
" been so impregnated with deleterious matter, that
" *not a fish is to be found in them,* whereas formerly
" those rivers were well stocked with fish. It is also
" shown in the same report, (page 20) that not only
" have fish been destroyed, but that "Animals grazing
" on the banks, *cows, horses, pigs, and poultry* have
" been *poisoned by eating the grass* which in times of
" flood has been covered by the infected waters."

" We do not wish to throw any obstacles in the way of
" trade. But we desire that the most efficient means
" should be provided for preventing *injury to the public,*
" and damages to *fisheries.*

" *We are confident that with care, and with compara-*
" *tively small expense, the nuisances which the public and*
" *the owners of fisheries so generally complain of may*
" *be prevented.*

" *Under the Gas Work Clauses Act,* 1847, *and under*
" *the Public Health Act,* 1858, *proprietors of gas works*
" *are subject to a penalty of* 200*l., and also to heavy*
" *daily penalties,* if they permit refuse from gas works to
" flow into any stream. We think that the occupiers of
" other works *should also be liable to* penalties to be
" *enforced in a summary manner for polluting public*
" *streams and the waters frequented by fish.* If there
" are any remedies under existing laws, *those remedies*
" *are so expensive,* that few will encounter the costs, and
" the promoters of nuisances relying upon this, go on
" undisturbed, *deriving benefit from the injury which*
" *they inflict upon the public.*

" Such then being the state of the case, *we entreat*
" *your lordship to lose no time in proposing such measures*
" as may seem best adapted to redress the injury com-
" plained of, and to *prevent the spread of this enormous*
" *evil.* Thousands of miles of streams which were de-
" signed by Providence to minister to the wants and
" necessities of man, give fertility to the earth, and
" beauty to the landscape, are not only rendered useless,

" offensive to the eye, and repugnant to taste and smell,
" but are changed into elements of disease and death.
" Where this state of things has existed for a long time,
" it cannot be remedied in a day. But no one who has
" paid attention to this subject has any doubt, that if a
" reasonable time be allowed for the carrying into effect
" of remedial measures, this can be done *without at all*
" *seriously interfering with the interests of trade and*
" *manufacture.*

" The country is greatly to blame for having permitted
" this evil to assume its present *gigantic proportions,* but
" if with our eyes fully opened, by the results of public
" enquiry, to its deleterious and demoralising influence,
" we permit it to *continue* and extend itself, we shall
" be unpardonable.

" But we need hardly remind your Lordship that every
" day's delay will increase our embarrassments, and that
" the sooner a check is put upon the present wholesale
" wanton destruction of one of the first necessaries of
" life, the easier will be our return to that system of
" effectual preservation of our streams from which un-
" happily we have so widely departed."

" We have, &c.

" (Signed)   EBURY,
" SHAFTESBURY,
" On behalf of the Sanitary Associations of
" Great Britain.

" SALTOUN, President,
" LLANOVER, Vice-President,
" Fisheries Preservation Association.

" To the Right Hon.
" Viscount Palmerston, K.G.,
" &c.   &c.   &c."

7th March,
1864.
3rd Annual
Report of the
Inspectors of
Salmon
Fisheries
(England
and Wales),
pp. 20, 21, 22.

In the 3rd Annual Report of the Inspectors of Salmon
Fisheries (1864), after instancing (from replies sent in by
Conservators of Rivers to the Inspectors, pages 20-22)
the *following rivers as variously polluted* by gas and dye-
works, sewerage, paper mills, lead and other mines, petro-
leum, tin, kianising, vitriol and creosote works, viz , the
*Eden, Kent, Derwent, Calder, Dee (at Chester), Dovey,
Tify, Rumney, Towey, Tave, Severn (at Gloucester),
Wye, Usk, Tamer. Teighn, Bovey, Exe, Wear, South
Tyne, and Tees,* and some of them *very grievously,* as

the *Calder, Dee* (at Chester), by gas and petroleum works; *Dovey, Towey, Tave, Tify, Wye,* and *South Tyne,* by lead and other mines, which have injured the Tify for fourteen or fifteen miles, and in the case of the *Dovey,* taken ten years' purchase from the value of every acre on its banks, and caused an injury to the neighbourhood estimated at £50,000; the *Usk,* by tin, gas, and iron works, vitriol, paper mills, tan and skin yards, and a creosote manufactory ; and the *Exe,* by gas, sewerage, paper mills, &c. The inspector (the late Mr. W. J. FFEN-NELL), remarks, page 22 :—

" The question of a remedy for pollutions is a very " large one."

" In the 1st Report of the Royal Commission on the " Sewage of Towns, 1858, allusion is made to the pol-" lution of rivers in the following terms, p. 11 :—

" ' Other evils of a less public, but still important " nature are caused by the pollution of watercourses " by town sewage. Even in the absence of large towns " below the outfalls, many small villages, &c., are " situated on the banks of streams. When such streams " are largely polluted by sewage, the comfort and " health of the inhabitants are interfered with, and the " value of their properties greatly deteriorated.

" ' The *destruction of fish* is another and very im-" portant consequence of the conditions described.

" ' The salmon fisheries of Scotland and Ireland not " only represent a large annual value, but they form " the occupations and livelihood of a very considerable " population. Apprehensions are already entertained " of serious injury by the daily increasing quantity of " sewage thrown into the rivers. Efforts have been " made with a view of arresting the evil ; and that it " can be arrested by means within our reach, is shown " in the case of Leicester."

Mr. FFENNELL then quotes from the same Report the conclusion arrived at by the Sewage Commissioners (before stated at page 7), and here repeated, viz. :—

" That the increasing pollution of the rivers and streams " of the country is an evil of *national importance* which

" *urgently* demands the application of remedial measures ;
" that the discharge of sewage and noxious refuse into
" them is a source of nuisance and *danger to health ;*
" that it acts injuriously not only in the locality where it
" occurs, but also on the populations through which the
" polluted rivers flow ; that it poisons the water which,
" in many cases, forms the *sole supply of the population*
" *for all purposes, including drinking ;* and that it de-
" *stroys the fish.*"

Session,1864.
Recommen-
dation of
Committee of
the House of
Commons, on
Sewage of
Towns.

In 1864 a Committee of the House of Commons recommended—

" That the important object of completely freeing the
" entire basins of rivers from pollution, should be rendered
" possible by general legislative enactment."

And the same Committee reported—

" In favour of the practicability of utilizing sewage by
" applying the same in the cultivation of the soil."

See Lord R.
Montagu's
Speech, 8th
March, 1865,
on the "River
Waters Pro-
tection" Bill.
—*Hansard,*
3rd Series,
vol. 177,
p. 1310.

In this year, and early in 1865, many towns memo-
rialized the Government to carry into effect the Committee's recommendations, among others *Nottingham, Shef*
field, Birmingham, Manchester, Preston, Coventry, Derby,
Wolverhampton, Bath, Huddersfield, York, Stockport,
Cheltenham, and *Oxford.* The memorials, &c. (or extracts from them) of *Sheffield, Nottingham,* of the
Rotherham and *Kimberworth* Board of Health, and of
Birmingham and *York,* are as follow :—

12 Oct. 1864.
Memorial of
Borough of
Sheffield,
[Parl. Paper,
6 March,
1865, page 5,
No. 105,] to
the Home
Secretary.

" Memorial of the Mayor, Aldermen, and Burgesses of
" *Sheffield* in Council assembled.

" Sheweth,

" That the practice of *discharging sewerage and other*
" *foul matters* into streams and rivers is *productive of*
" *great injury to the health of the people,* in consequence
" of the *pollution of the water.*
" That this sewerage may be converted into a permanent
" and increasing source of agricultural fertility by being
" conducted upon the land.
" That although it is a *nuisance at common law* to
" discharge any sewerage into rivers, yet *the law is in-*
" *operative* from various causes.

" That a Committee of the Honourable the House of
" Commons inquired into this subject during the last
" Session of Parliament, and recommended in their
" report thereon ' that *the important object of completely*
" *freeing* the *entire basins* of *rivers* from *pollution should*
" *be rendered possible by general legislative enactment.*'

" That your memorialists fully concur in this recom-
" mendation, and see no reason why the penalty for *dis-*
" *charging sewerage into rivers* should not be made as
" *simple and effective in application as the law now*
" *makes the penalty for injuries done to highways.*

" Your memorialists therefore pray that you will in-
" troduce a Bill in the next Session of Parliament to
" carry out *the recommendation of the Parliamentary*
" *Committee herein rehearsed, and effectually prohibit*
" *sewerage and any foul and offensive matter from*
" *being discharged into streams and rivers.*

" All which is respectfully submitted.

" Given under the Corporate Common Seal of the
" said Borough of Sheffield, this 12th day of
" October 1864.

" (Signed)   Thomas Jessop, Mayor."

At a meeting of the Sanitary Committee and other
Public Bodies of *Nottingham*, it was (*inter alia*) resolved :

" That in the opinion of this meeting the time has
" arrived when provision should be made by the Legisla-
" ture for enabling, and (when needful) requiring, the
" sewage of towns and villages to be applied for the
" benefit of adjacent districts of and, instead of polluting
" the waters of rivers and brooks, and also for enforcing,
" as far as practicable, that the noxious refuse arising
" from trade works be purified before it enters any stream
" of water, and the more solid parts of such refuse be
" separated and retained on the land."

" That the powers of the ' Public Health Act' and
" ' Local Government Act' do not completely meet the
" wants of the case, and they give no powers to restrain
" the *pollution of streams*, and cannot give that combined
" action over an extended area or watershed embracing
" several parishes, which is essential for providing an
" effectual remedy.

" That the neighbourhood of *Nottingham* shows the
" mischief resulting from the present state of things.

" That impure liquid matter from the manufactures
" and population of Old Lenton, near Nottingham, flows

8 Dec., 1864.
Resolution
passed at a
Meeting of
the Sanitary
Committee
and other
Public
Bodies of
" Notting-
ham," for-
warded to
Home
Secretary.
[Parl. Paper,
105,pp. 3 & 4.
6th March,
1865.]

B

"into the river Trent, about a mile and a half *above the*
"*point at which* A LARGE PART OF THE WATER SUPPLY
"*of Nottingham is now drawn from that river.*

"That the River Leen, which passes through this town,
"and which was *about 40 years ago a pure stream,* and
"afforded the principal supply of water to the town for
"all purposes, *is now foul and offensive by reason of its*
"*conveying part of the sewage of Nottingham, and the*
"*whole of the sewage of an extensive and populous higher*
"*district* over which the authorities of Nottingham have
"no control, and flows with *the rest of the sewage of Not-*
"*tingham,* into the parish of Sneinton, and *thence into the*
"*River Trent.*"

1864.
Memo-
rial of the
Rotherham
and Kimber-
worth Board
of Health to
the Home
Secretary.
[Parl. Paper
105, page 4,
6 Mar. 1865.]
Excessive
mortality of
Rotherham.

Memorial of the *Rotherham* and *Kimberworth* Board of
Health to the Home Secretary :—

"Sheweth,
"That this Board have been under deep concern on it
"appearing from the returns made by their officer of
"health from time to time, that the *mortality* of part of
"the district of the Rotherham and Kimberworth Local
"Board of Health (the town of Rotherham) *has been for*
"*some time greatly in excess* of the regular rates of mor-
"tality, having, for instance, in the two quarters ending
"*June 30,** been at the rate of *forty* in the 1,000!!

Epidemics
there in 1862
and 1863.

"That the town has on several occasions been subject
"to fatal epidemics, and in the *years* 1862 *and* 1863, *a*
"*medical officer from the Health Department of Her*
"*Majesty's Secretary of State* visited Rotherham to *in-*
"*quire into the state of its health,* and *especially with*
"*reference to the outbreak of typhoid fever.*

"That your memorialists believe the natural situation
"and state of Rotherham to be such as will not account
"for the sickness and death which have prevailed ; but
"they are *of opinion that being situate on the River Don,*

Memorialists
believe cause
thereof the
sewage
broughtdown
in river from
Sheffield.

"which flows from Sheffield, and brings down *an immense*
"*quantity of sewerage which falls into it at Sheffield,* and
"is deposited in the bed of the *river near Rotherham,*
"polluting the stream and poisoning the air, *is mainly*
"*the cause of the sickness and mortality which have pre-*
"*vailed, and which, to the belief of your memorialists,*
"*cannot be accounted for in any other way.*

"Your memorialists therefore pray that you will intro-
"duce a Bill next Session of Parliament, to carry out the
"recommendation of the Parliamentary Committee, that

* June 30, 1864.

" sewerage may be effectually prohibited from being dis-
" charged into rivers and streams.

(Signed)   " J. M. HABERSHON,
" Chairman of the Local Board."

From Memorial of the Mayor, Aldermen, &c., of *Birmingham* :—

" That your memorialists have been advised by the
" most eminent chemists and engineers on their difficulties
" in relation to sewage, and they have expended large sums
" of money and exhausted all their efforts in vain attempts
" to obviate the evils arising from it; and *they are now
" convinced beyond a remaining doubt, that the time has
" arrived for the introduction, by Her Majesty's Govern-
" ment, of a practical and comprehensive measure,* by
" means of which your memorialists may be enabled to
" carry the whole of their sewage, both liquid and solid,
" upon some adjacent lands, so that it may be applied, in
" accordance with natural laws, in adding to the fertility
" of the soil.

" Your memorialists hardly think it necessary to point
" out to Her Majesty's Government the *extreme impor-
" tance of preserving the purity of the rivers* and streams
" of this kingdom ; but they would respectfully suggest
" that the great and increasing number of towns and
" populous places exercising the drainage powers of the
" ' Local Government Act,' and other Acts of Parliament,
" in all parts of the kingdom, will *result in the intersec-
" tion of the Island* in all directions with a network of *open
" and noxious sewers* instead of the former pure and whole-
" some streams, *unless the evils arising from the present
" method of disposing of sewage are immediately arrested.*

" That your memorialists would also respectfully draw
" your attention to the *increasing difficulty now expe-
" rienced in obtaining a supply of water for large popu-
" lations from a pure and wholesome source ; because the
" rivers and streams are all becoming more and more in-
" infected with the pollution of sewage.* That your
" memorialists are surrounded by the very large popu-
" lations inhabiting the manufacturing districts of South
" Staffordshire and East Worcestershire, immediately
" adjoining the borough boundaries, being only separated
" from them by small streams, some of which, by means
" of the sewage of such populations, have been long since
" converted into *open sewers of the worst description,* and
" others are rapidly becoming in a similar condition.

1864.
Memorial of
the Mayor,
Aldermen,
&c., of
Birmingham
to the Home
Secretary.
[Parl. Paper,
6th March,
1865,
No. 105,
page 2.]

" Wherefore your memorialists urgently submit that it
" is *absolutely necessary that a Bill should be forthwith*
" *prepared under the direction of Her Majesty's Govern-*
" *ment*, and submitted to Parliament early in the ensuing
" Session, for enabling your memorialists, and other local
" authorities similarly situated, to accomplish the very
" important objects herein set forth.

 " Given under the corporate common seal of the
  " Mayor, Aldermen, and Burgesses of the
  " Borough of *Birmingham*, the day of
  " , 1864."

2 Jan., 1865.
Memorial of
the City of
York to the
Home
Secretary.
[Parl. Paper,
105, pp. 5 & 6,
6 Mar. 1865.]

" The Memorial of the Mayor, Aldermen, and Citizens of
 " the City of York, the Local Board of Health of
 " and for the same City,

 " Sheweth,

" That your memorialists regard the present mode of
" disposing of the sewage of cities and towns as *highly*
" *unsatisfactory, whether as regards the public health*
" or the economy of natural products applicable to the
" fertilization of the soil.

" That the pollution of the rivers and streams of the
" country by the discharge therein of the sewage of
" adjacent towns is productive of great and increasing
" evils, by *rendering the waters of such rivers and streams*
" *unfit for human consumption*, and converting what is
" *often the sole water-supply of a town into the fruitful*
" *source of disease and death.*

" That a Committee of the House of Commons reported

Session 1864.

" in the last Session of Parliament in favour of the practi-
" cability of utilizing such sewage by applying the same
" in the cultivation of the soil.

" Your memorialists therefore respectfully request that
" Her Majesty's Government will be pleased to introduce
" such a measure in the next Session of Parliament.

 " Given under our Common Seal, at the Guildhall
 " of and in the said City, this 2nd day of Janu-
 " ary, 1865.
  " (Signed) EDWIN WADE, Mayor."

January 1865.
Report of the
Special Com-
missioners on
Salmon Fish-
eries (Ire-
land).

In January, 1865, the Special Commissioners on the
Irish Salmon Fisheries, in their report for the year 1864,
at page 17, mention that " *the Liffey is fearfully pol-*
" *luted by sewage, which at certain places caused instant*

" *death to the fish.*" Also, that " the *Fisheries were*
" *largely injured by the water used in steeping flax,* the
" manufacture of which was greatly extending in Ireland."

The 3rd and final Report of the Commissioners on the
sewage of towns, 1865, says :—

  " As the result of our labours extending over eight
" years we have confidence in submitting to your Lord-
" ships the following conclusion :—

  " That the *right way to dispose of town sewage* is to
" *apply it continuously to land,* and it is only by such
" application that the *pollution of rivers can be avoided.*

  " We further beg leave to express that in our judg-
" ment the following two principles are established for
" legislative application :—

   " ' 1st. That wherever rivers are polluted by a dis-
" charge of town sewage into them the towns may
" reasonably be required to desist from further
" causing that public nuisance.

   " ' 2nd. That where town populations are injured
" or endangered in health by a retention of cesspool
" matter, the same may reasonably be required to
" provide a system of sewers for its removal.'

  " And should the law be found insufficient to enable
" towns to take land for sewage application, it would in
" our opinion be expedient that the Legislature should
" give them powers for that purpose."

30 Mar. 1865.
Third Report
of the Com-
missioners on
the Sewage
of Towns to
the Lords of
the Treasury.

To this Third Report of the Sewage Commissioners is
appended a most elaborate report made in 1864 by Dr.
Stevenson Macadam, F.R.S.E., &c., &c., on the hideous
contamination of the Water of Leith by the sewage of
Edinburgh and Leith, in which it is stated that :—

  " Into this small stream is discharged the sewage of
" 70,000 of the inhabitants of Edinburgh, and upwards of
" 30,000 of the people of Leith, and the result has been
" that the Water of Leith has become a foul polluted
" stream, conveying matter of the most disgusting and
" abominable character, and evolving fetid emanations
" into the surrounding atmosphere.

Page 6.
App. 5.

  " That the inhabitants of the districts bordering on the
" water complained bitterly of the offensive odours from
" the water, and which gave *rise to nausea and sickness,*

Page 8.

" and compelled them to keep their doors and windows
" shut.

"That Professor Simpson (now Sir James Simpson, Bart.,
" M.D.), showed from the *mortality* in the streets border-
" ing on the river, as compared with that away from its
" banks, that there was a *greater death rate* in the imme-
" diate neighbourhood of the Water of Leith than at a
" short distance therefrom.

" Thus taking a similar class of houses in the *Edin-*
" *burgh* district, and judging by the mortality among
" children under five years of age, Professor Simpson
" found that in the streets away from the influence of the
" foul water the mortality was in the proportion of 100,
" while in the streets near the Water of Leith the mor-
" tality was as high as 160!! In the *Leith* district also
" the death rate was greater, as in the streets *at some*
" *distance* from the harbour the mortality was in the
" proportion of 100, with a death rate among children
" under five years old of 1 in 12, while in the same class
" of streets *near the river* and harbour the mortality was
" 141, and the death rate among children 1 in 7!!

" That these statistics are positive evidence of the
" effects of the foul state of the Water of Leith conveying
" the sewage of Edinburgh and Leith, and the results
" are supported by the concurrent testimony of many
" persons who speak to the nausea and sickness brought
" on by the gases and vapours evolved from the water,
" and to the general ill health connected therewith.

Page 24.   As regards the *atmosphere* near the Water of Leith.

" The state of the atmosphere was not only judged of by
" the test of the nose but special experiments were made."

" Thirty-one samples of air were collected at various
" parts on different occasions. On the 7th April nine
" samples were tested, and whilst the degrees of purity
" of the air at three stations in Edinburgh away from
" the influence of the Water of Leith were respectively
" (100 being absolute purity) 85, 70, and 67, and the air
" at the Water of Leith at Coltbridge before being
" mingled with sewage was 75, the atmosphere in the
" immediate vicinity of the sewers and of the Water of
" Leith conveying sewage had its degree of purity re-
" duced to 63, 58, 55, and 55, and in one instance, as
" below the dam under the Water of Leith village, the

" 100 of standard colour was totally destroyed, a second
" 100 was similarly bleached, and of a third 100 only
" 20 remained."

On the 9th of April 16 samples of air were examined. Page 26.

" Three samples taken in Edinburgh in places away
" from the Water of Leith, and one sample collected in
" Leith at a distance from the polluted stream, gave
" respectively the degrees of purity 80, 75, 80, and 80,
" and one sample taken from the harbour at the Victoria
" Dockhead gave 70, while the air collected under the
" immediate influence of the Water of Leith conveying the
" sewage of Edinburgh and Leith gave respectively 60,
" 60, 50, 60, 60, 55, 55, 55, 60, 50, and 55.

" On the 14th April six samples of air were collected
" and examined, when it was found that over the Water
" of Leith before mixture with sewage the degree of
" purity was 80, while over the sewers and the Water of
" Leith conveying sewage the degrees of purity were 68,
" 66, 70, 64, and 70.

" In the whole course of the Water of Leith, from Page 27.
" Coltbridge downwards, not a single fish could be seen."

The Water of Leith at Edinburgh.

" The condition of the Thames at London is much less Page 30.
" foul than the water of Leith as it traverses Edinburgh.

" It will thus be observed that the *Water of Leith* as it Page 33.
" leaves Edinburgh contains fully *ten times* the quantity
" of organic matter which is found in the *Thames at*
" *London Bridge*, and necessarily the offensiveness of the
" water must be correspondingly greater."

Extract from Mr. FFENNELL's 4th Report, 1865, as 30th March
regards the pollution of Streams.   1865.
                                                  Inspectors of
Page 14 *of Report.*—" Public attention is now so ear- Salmon
" nestly directed to this question, and public opinion so Fourth Re-
" strong in *regard to the necessity of mitigating the evil* port.
" *complained of*, that it may not be in vain *to hope that*
" *some comprehensive measure may ere long be taken to*
" *abate a nuisance so excessive in its baneful effects*, in
" many ways as to alarm the minds of reflecting persons
" who are thoughtful and watchful of *the sanitary con-*
" *dition* of the people, and to create apprehension that it
" is insidiously in a less apparent manner generating

" disease in many districts, and *imperilling the general* " *health of the inhabitants of the country.*"

Adverting, page 28, to the cases of the *Rhiedol*, *Ystwith*, and *Dovey*, and observing that " no change for the better had been made in the condition of these rivers," polluted enormously by lead mines (the two first " completely poisoned,") Mr. Ffennell remarks, with reference to the Dovey, so seriously injured by the Dylifa lead mine, and the remark equally applies, he says, to *copper* mines, that—

" The managers of the Devon Great Consols mine have " shewn that the largest and richest mine in the kingdom " *can be worked without damage to the Fisheries,* and the " system pursued at that mine should be universally " carried out."

At pages 29 and 30 Mr. Ffennell notices some naptha and oil works as very destructive to the fish, the former near Gloucester, which were said to have " *poisoned the salmon last summer in great numbers,* and the latter below Chester, which so polluted the Dee, that it was said its water " could not be used for washing," it being added, that *scores upon scores of salmon had been found dead near the works,* and that *the water appeared at times blackened for miles.*

At *Page* 27 *of Report,* Mr. Eden, the other Inspector, says :—" *It cannot be too often shown that in most in-* " *stances the mischief occasioned by the pollution of rivers* " *is capable of easy remedy, and in all of great pallia-* " *tion;*" adding, at

Page 40 *of Report* :—" On the subject of pollution, I " have not suggested any amendment (Mr. EDEN refers " to certain amendments suggested by him in the English " Salmon Fisheries Act of 1861). It is *a question of* " *vital importance, not only to the Fisheries, but to the* " *health and enjoyment of the whole population of the* " *country, and appears to me to require graver considera-* " *tion* and more radical treatment than it can receive by " the insertion and discussion of a clause in a Fishery " Bill."

The Commissioners on the pollution of rivers in their 1st Report, 1866 (the Thames), state :—

20th March, 1866. 1st Report of the Royal Commission on the Pollution of Rivers, (the Thames).

" That *throughout the whole course of the river from* " *Cricklade to the point where the Metropolitan sewage* " *commences, fouling of the water* by sewage from cities, " towns, villages, and single houses, *generally prevails.* " The refuse from paper mills, tanneries, &c., passes into " the stream. There is no form of scavenging practised " for the surface water of the Thames, but carcases of " animals float down the stream until wasted by corrup- " tion. The river water receives unchecked the whole " of the pollution, solid and fluid, of the district ; and " this same water, after it has been so polluted, is ab- " stracted, sand-filtered, and *pumped into the Metropolis* " *for domestic uses.*"

Through its whole course to where Metropolitan sewagebegins the Thames fouled by sewage, &c.

Having described in much detail, pages 15 to 17, the enormous pollution of the Upper Thames by the sewage of Oxford, Reading, Windsor, Eton, Richmond, and Kingston, &c., &c., the report proceeds, page 17, thus :—

Towns of Upper Thames polluting the river by sewage.

" The *river basin at Hampton* (the pumping station of " the water companies) comprises an area of about 3,676 " square miles, and a population in 1861 of nearly " 900,000 *persons.* After a full allowance for retention " in cesspools, and for villages, &c., removed from the " banks of the river and its tributaries, there is no doubt " that the number of persons, whose sewage daily finds its " way into the water *from which London draws its supply,* " amounts to hundreds of thousands, and *this number* is " destined *greatly to increase* by the growth of popula- " tion, and by the development of the sewerage system " now only in very partial operation."

Sewage of hundreds of of thousands of persons finds its way into the water whence London draws its supply.

Page 18.—*Sir B. Brodie's evidence is conclusive,* that " there is no sufficient guarantee for its (the Thames' " water) arriving at Hampton purged of injurious taint. " The *London drinker of it* may be drinking with it *some* " *remnant of the filth of Oxford.*" " It is the general opinion of medical men, that what " causes the presence of organic matter in water to be " poisonous, is not its quantity but quality, and this " quality cannot as yet be detected by microscopic or " chemical analysis, and is indeed known only by its " occasionally noxious effects. The result seems to be,

Sir B. Brodie's evidence that the London drinker may drink some remnant of the filth of Oxford.

" that as a water supply the Thames, polluted with the
" sewage of the inhabitants of the River Basin, is open in
" kind, if not in degree, to the same objections as well-
" water infiltrated by liquid from an adjoining cesspool ;
" well-water which is so tainted beyond all doubt is liable
" to become poisonous. *Considering the enormous mag-*
" *nitude of the interests at stake in this question of the*
" *Metropolitan water supply* (the healths of many hun-
" dreds of thousands of persons), *it seems impossible to*

Only safe course is to keep sewage out of the water.

" *come to any other conclusion than that the only safe*
" *course is to keep sewage out of the river.* Each town
" needs to be protected from the abuses of towns above
" it, and to be prevented from committing abuse towards
" towns below.

" The question of sewage pollution of a river is an indi-
" visible one for the whole River Basin. Attempts to keep
" the main stream pure will be vain so long as tributaries
" are allowed to remain foul."

On the subject of disposing of town sewage, the Commissioners state that they fully concur with the Commissioners appointed "to inquire into the best mode of
" distributing the sewage of towns, and applying it to
" beneficial and profitable uses," who, in their final report
delivered in March, 1865, gave it as their unanimous

Right way to dispose of sewage is to apply it to land.

opinion, *after an investigation extending over eight years,*
that " the right way to dispose of town sewage *is to apply*
" *it continuously to land,* and that *it is only by such appli-*
" *cation that the pollution of rivers can be avoided ;*" and

And that wherever that application was in operation it was unattended by injury to public health.

they add, that such application of town sewage to land,
wherever that system is in operation, as at Croydon, Nor-
wood, Worthing, Carlisle, and Edinburgh, &c., was unattended by any injury to the public health.

And after various recommendations which the Commis-

Recommend that no sewage unless purified be cast into the Thames under penalties.

sioners humbly submit to Her Majesty respecting the
government and conservancy of the river, they recommend—

" That after the lapse of a period to be allowed for the
" alteration of existing arrangements, it be made unlawful
" for any sewage, unless the same has been passed over
" land, so as to become purified, or for any injurious re-

" fuse from paper mills, tanneries, and other works, to be
" cast into the Thames between Cricklade and the com-
" mencement of the Metropolitan sewerage system, and
" that any person offending in this respect be made liable
" to penalties to be recovered summarily."

On the 2nd May, 1866, Mr. Ffennell, the Inspector of Salmon Fisheries, presented his fifth annual report for England and Wales.

2 May, 1866.
5th Annual
Report
of Inspector
of Salmon
Fisheries.

On the subject of pollution by *collieries* and *paper mills*, Mr. Ffennell, at page 15, says :—

" I do not think I can better conclude my report than
" by giving an extract from the proceedings of the Wear
" Angling Association."

The extract, which conclusively shows that owners of *collieries* and *paper mills can* carry on their works without polluting the stream, is then appended. A portion of it, relating to collieries, is as follows :—

" The Earl of Durham had nobly led the way in reform
" by not only constructing subsiding ponds at all his col-
" lieries, but had in addition made staples or wells, into
" which the partially purified water was poured, thence
" pumped back again to the coal-washing apparatus, and
" so used over and over again *ad infinitum. By this
" simple plan being adopted, it became unnecessary to
" return a drop of foul water to the river or its tributa-
" ries."*

After stating that the example of Lord Durham had been or was about to be cordially adopted by Lord Vane and other large colliery owners, the extract says, with respect to *paper mills* :—

" Much complaint having been made as to the foul
" state of the *Browney*, arising from the flow of chemicals
" into the stream, we applied for information to Mr.
" Trotter Cranstown, of Churnside, who has large paper
" works on the *Whitadder*, in Berwickshire. His reply
" was as follows : · In reply to yours of the 20th instant,
" ' wishing for information as to the steps we have
" ' adopted to purify the waste ley from our paper manu-
" ' factory, we made a large pond, comprising nearly an
" ' acre of ground, which, being all sand and gravel

" ' below, formed a natural filter. It has proved
" ' thoroughly successful in practice, as *all the objection-*
" ' *able ingredients which polluted the river are kept back.*
" ' We also erected a large iron tank, whence all the
" ' strong ley is pumped up and re-used, so as to form an
" ' actual saving.' "

The extract concludes thus :—

" *Experience has also satisfied us, that with trifling*
" *exceptions, the collieries and gasworks, paper mills and*
" *manufactories, can be carried forward equally well*
" *without fouling our once pellucid streams.* In nearly
" every case it is *but the outlay of a little extra capital,*
" and when great concerns are planted near, and have the
" use of our streams for their commercial purposes, surely
" it is not asking too much that the money advanced to
" plant a business should include the fractional sum neces-
" sary to prevent the owners of that concern from wilfully
" and unnecessarily destroying the property and rights of
" their neighbours."

August, 1866. Registrar General's weekly return.

In the supplement to the Registrar General's weekly return for August 6th, 1866, appear the following remarks on the *water supply* of the east districts of London, taken from Professor Frankland's Report :—

" The cause of the epidemic of Cholera consists, as is
" well known, of a zymotic matter in various degrees of
" activity, all over the London area, affecting the people
" in various ways through air, contact, and *water.*
" Hitherto in all great outbreaks here, the cholerine
" which this stuff may be called, has been distributed
" chiefly through *water.*
" It is the amount of *organic matter* contained in this
" *water* which is of *special importance* in connection with
the *outbreak of Cholera.*"

1866. Regr. Genl.'s Report of the public health.

From the Registrar General's Report on the public health, for the year 1866,

" Dr. Farr states that there is *no apparent decline* in the
" rate of deaths from fever   He considers it extremely
" probable that *typhoid fever* is sustained by the *increas-*
" *ing contamination of the waters.*"

6 May, 1867. 2nd Report of the Commissioners on the Pollution of Rivers. (The *Lea.*)

On the 6th May, 1867, the Commissioners on the Pollution of Rivers presented their 2nd Report which refers to the *Lea.*

It shows (pages 11 to 13) that the *Lea*, from which water *for the domestic use of a large portion of London* is taken, is polluted from *Luton* (close to its source) to West Ham, near Blackwall (where its mouth is), by all the towns and places on it, by the sewage of *Luton*, numbering 20,000 inhabitants and upwards, and by its manufacturing refuse (from the preparation of straw plait), composed of large quantities of metallic salts, dye stuffs, brimstone, &c., and in some cases *poisonous* materials, by the sewage of *Hatfield*, which, though not discharged directly into the river, the Commissioners say finds its way there, by the sewage of Whitwell and Welwyn, where it is also polluted by *arsenic* which comes from the wool of sheep when they are washed, and which has been retained in their wool since the previous "dipping," in which process *arsenic* is used; by the sewage of *Hertford*, which, though passing into the river in a mitigated form, is a constant source of complaint to *Ware*, situate below; by the sewage of *Ware* itself; and by the sewage respectively of *Bishopstortford* (*via* a tributary of the *Lea*), *Hoddesdon, Broxbourne, Cheshunt, Waltham Abbey, Waltham Cross,* and *Enfield-Highway.* All the foregoing *pollution* taking place *above the intake of the East London Water Company.* Below the intake, the report states, the *Lea* receives the sewage (unmitigated by any process whatever) of *Enfield, Edmonton, Hornsey, Chipping Barnet, East Barnet,* and *Hadley,* and also of *Leyton, Leytonstone, Walthamstow,* and *West Ham,* whose population (*West Ham*) alone at the last census was 22,337. The report adds that this pollution of sewage was much on the increase, that the river at *Old Ford* was rendered very PESTIFEROUS during hot weather by impurities from chemical and other works, and that the district bordering on the tidal portion of the *Lea* has become a nuisance district, the seat of trades expelled from the better parts of the metropolis.

The Commissioners, after condemning the East London

Water Company for having drawn " unfiltered water for
" domestic consumption " during July and August, 1866,
from the Old Ford reservoir, to which fact had been
attributed the outbreak (or great aggravation of the outbreak) of cholera in the East of London of that year,
state (p. 26) as one of the conclusions they had come to—

" That it is expedient that more stringent measures be
" adopted to protect from pollution that portion of the
" Metropolitan water supply which is derived from the
" Lea."

And, finally, the Commissioners as in the case of the
Upper Thames recommended (among other recommendations)—

" That after the lapse of a period to be allowed for
" alteration of existing arrangements it be made unlawful
" for any sewage, unless the same has been passed over
" land so as to become purified, or for any injurious
" refuse from manufactures or agriculture to be cast into
" the river Lea, or into any of its tributaries, and that
" persons offending in this respect be made liable to
" penalties to be recovered summarily."

15 July, 1867.
6th Annual
Report of In-
spectors of
Salmon Fish-
eries, (Eng-
land and
Wales).

In July, 1867, the Inspectors of Salmon Fisheries
(Mr. Frank Buckland and Mr. Spencer Walpole) presented their 6th Annual Report for England and Wales.

Mr. Frank Buckland, after stating that his predecessor
(the late Mr. Ffennell) had issued a series of questions to
the Boards of Conservators of rivers, of which question
No. 12 related to "pollutions," gives in an Appendix at
pages 38 to 61 the answers received by the Inspectors to
that question.

From these answers it appears that of 20 rivers named
in the Inspector's 3rd Report, 1864, (page 14, ante) as more
or less polluted, 16 continued to be polluted at fully the same
degree, (especially the Dovey and its tributary the Twymin
by the Dylliffa mine and by Sir John Conroy's), while
the names of about 16 additional rivers are given as suffering from lead mines and numerous forms of pollution, which

iu many instances the answers say *kill vast quantities of fish*, as in the cases of the *Allyn* (tributary of the Dee), " *where for* 14 *miles every fish in Dec.*, 1866, *was destroyed by petroleum works*,"* the *Usk*, of which it is remarked, it is so polluted that " unless *some legislative measure stop the evil the objects of the Salmon Fishery Acts and the labours of the Conservators will be thrown away*," the *Exe*, where paper mills " *kill bushels of fish ;*" the Trent, where (at Burton) for several years past, " *fish have been poisoned by tons ;*" the *Aire*, which is " so surcharged with immense " quantities of coal dust and dye, that *hundreds of salmon* " are choked and blind-folded by the poisonous salt ;" and the *Wear*, of which the answer says, " it is dreadfully pol- " luted by lead mines, collieries, iron, gas, and chemical " works, paper mills, sewerage, and every abomination a " thickly populated district can put into it."

This Report gives therefore a total of at least 32 rivers thus poisoned and polluted, very many of them horribly.

As to the pollutions by lead mines Mr. Buckland at page 4, says, " the measures taken to obviate this terrible evil have been but slight," he points however to the answer of the *Tamar*, page 54, where it is stated that—

" The Devon Great Consols Mine Company (whose " good example was noticed in Mr. Ffennell's 4th Report, " page 24, *ante*), have made catch pits, &c., and are

* Referring to this fearful case, Mr. Mostyn Owen, the honorary secretary of the Dee Fishery Board, told the great Salmon Fishery Congress at South Kensington, in June 1867, that a skilful analyst, after analysing the water, had declared to him that the river which was thus polluted, and which supplied all Chester with its drinking water, " *might any day bring down such a quantity of deadly poison* " *that half Chester* MIGHT BE KILLED BY IT ;" Mr. Owen adding that the Fishery Board had done what they could to obtain a conviction against the offenders, but had failed from a mere technicality. And in reference to this same river, Lord Robert Montagu informed the House of Commons, in the course of his speech in moving the second reading of the " River Waters Protection " Bill in 1865, that " the " honorary secretary of the Dee Fishery Association had preserved a " bottle of *pure paraffin* made from the water taken from the Dee " below the petroleum works."—*Hansard*, vol. 177, 3rd Series, page 1316.

" saving the arsenic with profit to themselves. No other
" mining company has used any effectual measures." *

And speaking of polluted water generally, Mr. Buckland remarks, page 5:—

" Impure and polluted water will encourage disease,
" especially cholera: pure water will disarm disease of its
" powers, and at the same time be available for growing
" excellent human food."

Mr. Walpole the other Inspector, in his separate Report
of the same date observes, page 5:—

" It would be impossible for me to omit all notice of
" that bane of Salmon rivers pollutions—a bane which
" unhappily there is but a very doubtful remedy against
" under the existing laws;" adding, "but though I am
" aware the law is defective, and though I hope the day
" may shortly come when it may be made illegal to put
" any pollution into a running stream, I abstain from re- .
" commending any alteration in the Salmon Laws in this
" respect as a question of such importance should not be
" confined to the minor consideration, important though
" it be, of the cultivation of our Salmon rivers."

15 Aug. 1867.
3rd Report of
the Commis-
sioners on the
Pollution of
Rivers. (The
Aire and
Calder.)

On the 15th August, 1867, the Royal Commissioners on
the pollution of rivers made their Report on the Rivers
Aire and *Calder*. At pages 10 and 11 the report, says:

" The *Aire* and *Calder*, throughout their whole course,
" are abused, obstructed, and polluted (to an extent
" scarcely conceivable by other than eye-witnesses) from
" *Skipton* on the *Aire*, and from *Todmorden* on the *Calder*,
" to *Castleford*," where they unite.

* Except, it is only right to add, (but of which the *Tamar* Board
were probably unaware,) the *London Lead Company* and *Mr. Wentworth Beaumont, M.P.*, both of whom, to their credit, had adopted,
or were about to adopt, at their extensive mines in the valleys of the
Wear and *Tees*, the same means as the Devon Great Consols Company to prevent the pollution of the rivers.—(*Vide* correspondence
between the London Lead Company and Mr. Beaumont and the late
Mr. Ffennell, the Inspector of Fisheries. *Land and Water*, 13th
October, 18(6, p. 276.—Also, Mr. Buckland's remarks on Mr. Beaumont's good example, pp. 14 and 49 of the Inspectors of Fisheries
Seventh Annual Report, 1868.

" Our inspection was corroborated by uncontested and
" overwhelming evidence."

" The rivers *Aire* and *Calder* and their tributaries are
" abused by passing into them hundreds of thousands of
" tons per annum of ashes, clay, and cinders from steam-
" boiler furnaces, ironworks, and domestic fires ; by their
" being made the receptacle, to a vast extent, of broken
" pottery and worn-out utensils of metal, refuse from
" brickyards, &c., earth, stone, &c., from quarries and
" excavations, road scrapings, street sweepings, &c.; by
" spent dye-woods and other solids used in the treatment
" of worsteds and woollens ; by hundreds of carcases of
" dogs, cats, pigs, &c., which are allowed to float on the
" surface of the streams or putrify on their banks ; and
" by the flowing in, to the amount of *very many millions*
" *of gallons per day*, of water, poisoned, corrupted, and
" clogged by refuse from mines, chemical works, dyeing,
" scouring, and fulling worsted and woollen stuffs, skin-
" cleaning and tanning, slaughter-house garbage, and the
" sewage of towns and houses."

" Many streams where, by reason of their foulness, no
" form of life can at present be found, persons now living
" recollect abounding in fish."

One enormous penalty paid for this abuse of the rivers
is flooding, consequent on the raising of the rivers' beds,
and at page 12 the Commissioners exemplify this as
follows :—

" That on the 15th of November, 1866, rain com-
" menced and continued for several days, flooding the
" valleys of the *Aire* and *Calder* most destructively from
" the mountains to the sea. In several instances persons
" were washed away and drowned. It is not possible to
" form an estimate of the money value of the damage
" caused to the manufacturers, landowners, and others in
" the West Riding.
" The *total loss* was locally estimated at *from half a*
" *million to a million sterling*.
" The lesser sum would have been sufficient to put the
" rivers in a condition to render such destruction of life
" and property impossible."

 After stating at page 13 that the amount of solids taken
into the streams from sewers is in the aggregate enormous,

c

and that at *Leeds* the entire volume of sewage of *eight to ten million gallons* per day passes into the *Aire*, as also that of *Bradford*, *Keighley*, and *Skipton*, and that the *Calder* receives all the sewage of *Todmorden*, *Halifax*, *Huddersfield*, *Dewsbury*, *Wakefield*, and of smaller towns, &c., &c., the Commissioners then declare—

" That the present gross abuse of the rivers we " inspected *may be in a great measure prevented, and in* " *such manner and at such cost as to be beneficial to all* " *parties.*"

And speaking of the woollen manufacture, they remark :

" In the two facts—first that in one year (1864) " 384,000,000 pounds of wool were worked up into " various tissues in Great Britain ; and secondly, that " *every pound* of this wool has to undergo operations " necessitating *the use of large volumes of water*, and " rendering *that water foul and offensive*, we have the " history of all rivers on which that trade is located, and " *notably* of the *Aire* and *Calder.*"

·And they add—

" That with very few exceptions the streams of the " West Riding run with a liquid more resembling *ink* " than water."

Referring to the tanneries at Leeds, the report states (page 35)—

" That as many as 2,750,000 hides were annually con- " verted into leather in Leeds and its neighbourhood."

And at page 36, speaking of the pollution of the river by tanning refuse, the Commissioners say—

" We believe that the pollution of the river, which is " undoubtedly very considerable from tanning refuse, " *may be prevented without injury to this very important* " *industry* " (the leather trade).

They then give a detailed description of the condition

of each town visited by them on the *Aire* and *Calder*. Of *Skipton* they say (page 37) : —

"Its sewage and refuse of its dye-works, slaughter-
"houses, tan-works, paper-mills, and factories, all go into
"the river." Of *Keighley :* "That the sewage is the
"*cause of great nuisance*, and that the *workmen* at the
"water-wheeels *have frequently to leave the mills* sick*."
Also "that soap-suds flow into the river, which ferment
"and give off large volumes of foul gases, and that *local
"disease* is imputed to this form of pollution."

And much the same story is told by them of *Bingley*, all its filth and refuse going into *Harden Beck* and into the *Aire*.

Of Bradford, with a population in 1861 of 48,646, they say :—

"The whole of the sewage of Bradford and af the
"populous district above the town flows into the *Beck*,
"producing an *indescribable state of pollution*. At the
"time of our inquiry Bradford Beck was the source of
"supply to the Bradford Canal, the fluid of which became
"so corrupt in summer that *large volumes of inflammable
"gases were given off*." And "though," the Commissioners add, "it has been considered an impossible feat to
"set the Thames on fire, it was found practicable to set
"the Bradford Canal on fire, as this at times formed the
"amusement of boys in the neighbourhood, the flames
"rising *six feet high*, and running along the surface of the
"water for *many yards*, like a will-o'-the-wisp, and *canal
"boats* have been so *enveloped in flames* as to frighten
"persons on board."

Proceeding (page 39) to describe the condition of the *Aire* at Leeds and the streams communicating with it there, the Report says :

"The entire volume of the sewage of Leeds, solids and
"fluids (from 7,000,000 to 12,000,000 gallons per day),
"is turned into the *Aire*, creating a great nuisance."
"The whole of the becks flowing through the town are
"fouled with waste refuse from dye-works, tanneries, and
"the various other manufactures, from their sources

---

* Of which there are 54.

c 2

" beyond the municipal boundaries to the *Aire,* which
" river is also polluted along both margins,* carcases of
" dead animals float down until intercepted by shoals,
" where they remain to become putrid. As the popu-
" lation of Leeds has increased and manufactures have
" been extended, the local death rate *has risen* until the
" town is ranked by the Registrar-General *amongst the*
" *most unhealthy of the kingdom.* The death rate from
" 1857 to 1860 was 28 per 1,000, and from 1860 to 1864
" was 29·5 per 1,000."

The Commissioners next describe, at pages 40 to 43,
the condition of the River *Calder* as affected by the
various towns on it or its tributary streams, stating that
all such towns as *Todmorden, Halifax, Huddersfield,
Dewsbury,* and *Wakefield,* with an aggregate population
of nearly 250,000, discharge their sewage and the filth
from their various manufactures into the *Calder,* Halifax
discharging the same by the *Hebble,* and Huddersfield by
the *Colne,* the pollution of the river caused by *Wakefield*
being much increased by a large weekly cattle and sheep
market which is held there, and where as many as 1,000
horned cattle and 13,000 sheep have been sold in one day.
" In dry weather," the Commissioners mention, " the
" banks of the river are covered with dark thick slime, and
" the entire volume of water is stained with dye-refuse ;
" and from a river thus polluted, *Wakefield,*" the Commissioners add (at page 46), " *draws three-fourths of its*
" *water supply.*" And at the same page and page 47
they declare " *that there is no excuse for the discharge of*
" *any sewage or ochrey matter into running streams and*
" *rivers, and that the pollutions from soap-suds, dyeing,*
" *and other processes may be prevented or greatly*
" *alleviated,*" the Commissioners concluding their detailed

* And further on at the same page (39) the Commissioners say :—
" *Carcases of pigs, cats, dogs,* &c., have been removed to the exten
" of *fifty in a day,* but it must not be inferred that they are removed
" with any degree of regularity."

account of the pollution of the river by these towns as follows :—

" The injury inflicted by the river pollution of these
" and other towns to the estates of many riparian owners
" is very great; streams which in the memory of men
" now living were comparatively pure and well stocked
" with fish are *now black and stinking.* The land through
" which such polluted streams flow is ruined for resi-
" dential purposes, and is injured and reduced in value.
" Even for mill purposes the water is so bad as to be con-
" sidered unfit for manufacturing uses, and other sites
" are selected, where water can be obtained from canals
" or by sinking wells. In many instances cattle will not
" drink the local river water, and farms are depreciated
" in consequence. Large country houses, which for-
" merly, with their river frontage, rights of fishing, and
" ornamental gardens, were valued as residences, have
" been abandoned, or are let merely at farming rents.
" The cattle plague prevailed to a great extent in the
" Thorpe Hall meadows below Leeds, and on lands
" bordering foul rivers in other districts. This fatal
" disease was considered by the tenants and pro-
" prietors to have been aggravated by the foul state of
" the water and by the tainted atmosphere, caused by
" river pollutions."
" In some cases the manufacture and dyeing of finer
" sorts of goods has necessarily been abandoned, and in
" other cases extension of manufacture is rendered impos-
" sible, because there is no additional clean water to be
" obtained in the district."

And at page 51 the Commissioners point out that

" One great argument for the purification of the rivers
" up to an available point should not be passed over in
" closing the consideration of this question. Manu-
" facturers require *clean* rather than *pure* water. They
" cannot get on with water fouled by solid suspended
" matter."

Having shewn (pages 51 to 53) that the existing laws
are practically, for one reason or another, quite insufficient
to correct the dreadful evil of the pollution of streams,

the Commissioners state (at pages 53 and 55) as their
conclusions—

" That in order to prevent the pollution and legally
" control the management of rivers, their basins or water-
" sheds must be placed under supervision, irrespective of
" any arbitrary divisions of county, parish, township,
" parliamentary, municipal, or Local Government Act
" boundaries, or indeed of any artificially established
" division."

" That the question of profit and loss in abating nui-
" sances and in preventing the pollution of the atmos-
" phere, running waters, or sea shore, by town and house
" sewage, or by working mines and carrying on trades
" and manufactures, ought not to be too rigidly taken
" into account."

" That the prohibition against the *casting in of solids*
" *may at once be general without any exception.*"

" That a stronger power than has hitherto been avail-
" able must be brought to bear if the present abuse and
" pollution of streams is to be arrested, and *Government*
" *supervision and inspection must enforce the action of*
" *local authorities.*"

" That our experience of the weakness inherent in
" unaided and uncontrolled local authorities convinces us
" *that a central board appointed by a State department is*
" *necessary to the efficient protection of running waters.*"

From the *Medical Times and Gazette* of 28th Sep-
tember and 23rd November, 1867, with extracts from Mr.
Simon's Annual Report to the Privy Council on the
Public Health, for 1867 :—

" A reduction of typhoid fever is undoubtedly shown
" coincident with sanitary improvements. ' Though not
" ' with absolute constancy, drying of the soil of a town
" ' and reduction in the crowding of houses have been
" ' followed by reduction of fever. Much *more important*
" ' appears to be the substitution of an *ample supply of*
" ' *good water* for a scanty and impure supply.' "

Mr. Simon insists with the utmost force upon the absolute relation of cholera in England to faults of drainage
and *water supply* :—

" It cannot be too distinctly understood," he says,

" that the diffusion of cholera among us depends entirely
" upon the numberless filthy facilities which are let exist,
" and specially in our larger towns, for the fouling of
" earth and air and *water*, and thus secondarily for the
" infection of man, with whatever contagium may be con-
" tained in the miscellaneous outflowings of the popula-
" tion."

" Dr. Buchanan's Report on Typhoid Fever
" at Guildford, in Sept. 1867.

" There have occurred some *local outbursts of fever*
" which, as especially the outbreak at *Guildford, have
" preached in terrible language*, in the startling tones of the
" angel of death, *the frightful and inevitable dangers of
" an impure water supply*, enforcing *on even the most care-
" less and ignorant* the paramount necessity of *unwearying
" and incessant vigilance against the possibility of the
" water supply being*, in any part of its course, *tainted by
" sewage.*

" Typhoid fever, it seems, is by no means uncommon
" in and about Guildford, and before the particular out-
" break several sporadic cases had occurred. But in the
" last three days of August cases of typhoid were ob-
" served in the more elevated and healthiest parts of the
" town. On September 3 and 4, 'a surprisingly large
" number of people sent for medical assistance; in the
" first ten days of September about 150 *cases* altogether
" had come under treatment, and by the end of the month
" that number had increased to 264. A remarkable
" feature of the outbreak was its localisation; and that,
" too, chiefly in the highest levels of the town, without
" distinction of social position and circumstances. The
" acmé of the disease's progress was reached before the
" middle of September, and it thence declined rapidly.
" Inquiring as to the causes which could have induced the
" disease, Dr. Buchanan soon came to the conclusion that
" drainage played no direct part in the matter; and the
" next point to ascertain was the *state of the water
" supply.*

" The public waterworks have their source in two wells
" sunk some twenty feet into the chalk at the lowest part
" of the town; one of these is an old well, from which
" water is raised by the power of an adjacent water-mill;
" the other is a new well, ' from which, for a short time
" in the middle of the present year, water was distributed

" Medical Times and Gazette." of 23rd Nov., 1867, on Report of Dr. Buchanan, of the Medical Department of the Privy Council, on the Fever at *Guildford* in Sept. 1867.

Attention directed to water supply.

" to the higher parts of the town by engine power,' but
" the pumping engine broke down on August 1st, after
" which date no water was drawn from the new well, the
" supply being from the old well to both the high and
" low services of the town till 17th August.

" On that day, the water which had been stored in the
" *new reservoir* (from the new well) was distributed to the
" *high-service* houses of the town : ' it was distributed on
" no other day, and to no other houses.' Now, it turns out
" that all the houses first attacked with typhoid fever in
" the beginning of September had received water from *this*
" reservoir ; and, without giving detailed statistics, it is
" sufficient to state that of the 150 cases occurring in the
" fortnight from August 28 to September 10 there were

Dr. Buchanan considers the outbreak was due to impure water supply.

" hardly a dozen persons attacked ' *who had not* had daily
" and hourly access to the water of the *high service*..' Dr.
" Buchanan was unable, on careful and detailed inquiry,
" to find any condition at all coincident with the outbreak
" of the fever, save that of the water supply.

Further source of impurity in water supply suggested, viz., the communication between Water Company's pipes and river (Wey), which is greatly contaminated

" But Dr. Buchanan suggests a possible further source
" of impurity in the water supply. ' *There exists a com-*
" *munication between the river and the pipes of the water-*
" *works. This is said to be very rarely used*, and only
" for the purpose of getting a first sucking power to the
" pumps ; and it is stated not to have been used at all
" during the present summer.' Whether this latter aver-
" ment be true or not, is immaterial now, but with the
" story of the East London cholera outbreak fresh in our
" remembrance, we *cannot but protest against the power*
" *thus left in the hands of the water engineer to throw at*
" *any moment into the service mains the water of a river*
" *so contaminated as the Wey at Guildford* must, of
" necessity, be.

" In *country towns* and outlying hamlets, the *poisoned*
" *water spring saps the vigour of the population, and*
" *swells, to a degree much greater than is generally sup-*
" *posed, the death-rate of the whole country.*"

8th February, 1868.
The *Lancet*, with extracts from Report of Dr. Thorne of the Medical Department of the Privy Council on the *Terling Fever* of 1867-8.

THE OUTBREAK OF FEVER AT TERLING, ESSEX.

" The lengthened inquiry of Dr. Thorne, the medical
" inspector sent down to Terling by the Medical Depart-
" ment of the Privy Council Office, has been brought
" to a close.

" The present epidemic is one of true typhoid fever,
" and deserves notice both on account of its magnitude

" and the suddenness of its occurrence ; for, says the
" reporter,—

" ' Almost entire families were attacked, in some
" instances the patient seems to be almost instantly
" overwhelmed with the intensity of the poison. . . On
" the 13th of January, 1868, 208 *persons* had already
" been attacked by the prevailing malady, and several
" fresh cases were daily occurring. I do not include in
" my number cases of so-called diarrhœa (possibly all
" mild cases of typhoid fever) which were only accidentally
" heard of, and which probably were numerous, though
" not of sufficient severity to call for medical relief."

" The first case of fever occurred on the 13th of
" November ; the next not until December 4th, when
" the epidemic seemed rapidly to develop, for the report
" says—

" ' During the following ten days 30 fresh cases were
" seen ; but on the 15th, 16th, and 17th by far the largest
" number were attacked, 22, 19, and 12 cases occurring
" respectively on those days. After this, though the
" daily number of fresh cases was by no means so large,
" still a steady increase took place. . . As yet, however,
" only one death had taken place—namely, on Dec. 14th ;
" but when the third week of the epidemic had arrived, a
" steadily and gradually increasing death-rate commenced,
" and on the 30th twelve of the patients had died, and
" others were dying. *Terling was now completely panic-*
" *stricken.* . . No class of persons was exempt ; the
" rich, the well fed and clad, were attacked in common
" with the poor and destitute. At Lord Rayleigh's
" residence 10 cases had occurred, the vicar's house was
" a seat of the epidemic, and from one end of the village
" to the other the disease seemed to be almost evenly
" spread. Age and sex seemed to present remarkable
" peculiarities : thus, out *of* 145 *cases* whose ages I was
" enabled to obtain, 79 were children under fourteen years
" of age, and of the remaining 66, 50 were females,
" leaving 16 males whose ages exceeded fourteen years,
" out of the entire number attacked."

" And this peculiarity as to the age and sex of the
" persons attacked the reporter expresses his belief is to
" be ascribed to the fact—

" ' That the men, and the majority of the boys over
" fourteen years of age, spend the greater portion of their
" time away from home, labouring in the fields, and that

Up to 13 Jan. 208 persons attacked and fresh cases occurring exclusive of milder cases of so-called diarrhœa probably numerous.

Terling panic-stricken on 30th Dec.

Fact of so few men and boys over 14 years of age being attacked attributed to their principally drinking beer.

" they principally drink beer ; *whereas the women and*
" *children are left at home,* and procure a *considerable*
" *portion of their beverage from the wells,* the *children*
" *drinking directly from them* much more frequently
" *than the women,* owing to the latter consuming a *good*
" *deal* of tea, in making *which the water is of course*
" *boiled.*

Population of Terling 900.

" The causes of this severe epidemic, which up to Jan.
" 13th, had stricken down 208 *of the inhabitants,* and
" caused upwards of 20* deaths in a population of 900 *souls,*
" were due, as we stated in our columns on the 18th of
" January last, to local influences, and these have been
" proved (as we then surmised they would be proved) to
" be the existence of overcrowding, the accumulation of
" filth and nuisances, and THE USE OF POLLUTED WATER."

The *Lancet* then quotes from Dr. Thorne's report, as an
illustration of the general state of things at Terling, a
passage showing that the whole filth of every kind, without any exception, of four cottages found its way into a well,
5 ft. deep, which supplied the inmates with their water.

From the Registrar-General's Report of Professor
Frankland's analysis of the water supplied to Londen
during 1867, showing an *increase* of solid impurity in the
water over 1866 :—

" *Professor Frankland* has reported on the waters sup-
" plied to the metropolis in the year 1867. He states
" that the search for *sewage pollution in the metropolitan*
" *waters* has now *assumed a high degree of importance.*
" In 1867 the total solid impurity exhibited *an increase*
" *over* 1866 in the waters supplied by the different com-
" panies, except in the case of the East London Company,
" in which a marked decrease had occurred."

Fever at Cardiff (Canton District). The *Lancet,* 8th Feb., 1868, with Report of Dr. Taylor to the Local Board of Health.

" FEVER AT CARDIFF (CANTON DISTRICT).

" Fever has been very prevalent of late in Canton, a
" district of Cardiff ; and no wonder, considering the posi-
" tively offensive *pollution* of air and *water* which exists
" there. A *very able and practical report on the sanitary*
" *state of the locality* has been *presented to the local board*
" *of health by* Dr. TAYLOR, and we regret that we cannot

* On March 14 the deaths had numbered in all at least 44.

" find space for a full notice of its details. *The water*
" *supply is not only scanty, but largely polluted by sew-*
" *age.* No wonder the death rate of Canton last year
" (27.28 per 1,000) was higher than that of Cardiff
" (23.54)."

" BAD WATER AND ITS INFLUENCE ON HEALTH,"

" A short time ago we directed attention to an outbreak
" of *typhoid fever in the Royal Marine Barracks at*
" *Stonehouse ;* * and in reference to it the *registrar of the*
" *district* reports :—

" ' The *water* which was used by the men from a well
" adjoining the new wing of the barracks is clear and
" sparkling, and apparently much better than that sup-
" plied by the Devonport Water Company. *In conse-*
" *quence of its being much in requisition*, there was a
" ': great drain from its source, which I have every *reason*
" *to believe is a very large natural cavern*—one of those
" which are so often found in the limestone formation.
" This cavern is upwards of 500 feet in length, and
" *over or on its sides about thirty houses have been built.*
" Persons who do not know of its existence have won-
" dered how it is that their cesspits have not required
" to be emptied ; *which, however, is no wonder to me,*
" *for I feel convinced that many of the cesspits drain*
" *into it.* When the foundation was making for the
" new wing of the barracks, I intimated to the clerk of
" the works the existence of a large cavern, running
" N.E. to S.W,; and it was ultimately cut into, showing
" its presence within about 50 feet from the top of this
" very well."

Marginal notes: Fever in Marine Barracks at Stonehouse (Plymouth). The *Lancet*, 8th Feb., 1868, with Extract from the Report of the Registrar of the district Water used supplied from a well. Believed to be polluted by cesspits draining into it.

From the *Lancet*, 8th February, 1868, giving Professor
Frankland's remarks on the water supply of London
during January, 1868 :—

" Professor Frankland's report to the Registrar-General
" on the quality of the *waters* supplied to the metropolis
" *during last month* is more than usually suggestive of
" the undesirability of rivers as sources whence to draw
" water for potable purposes. Heavy rains caused the
" Thames to overflow its banks above the points of intake

* Fatal to seven of the marines.

" of the London Water Companies ; in addition to the
" soluble matters washed down from cultivated fields,
" *great quantities of putrescent animal matters were*
" *flushed out of the sewers of Oxford, Windsor, &c.,*
" *into the river ;* ' hence the large proportion of organic
" ' carbon and nitrogen, and the great previous sewage-
" ' contamination in the waters of those Companies which
" ' derive their supply from that river.' In fact, it is
" shown that the Thames was contaminated, at the time
" the samples for Dr. Frankland's analyses were taken,
" ' *with double the usual proportion of organic matter.*'
" The waters delivered during the latter part of the month
" were ' in *such a muddy condition as to render them*
" ' *totally unfit for domestic use.*' "

LONDON WATER.

From the Registrar-General's Return, week ending
Feb. 29th, 1868 :—

" The results of Dr. Frankland's analyses of the waters
" derived from the *Thames* are *not satisfactory.* The
" waters contained impurities probably of an animal
" origin to a considerable extent."

23rd March, 1868. 7th Annual Report of Inspectors of Salmon Fisheries (England and Wales).

On the 23rd of March last the Inspectors of Salmon Fisheries presented their seventh annual Report.

Forty-nine rivers are reported on, namely, the *Aln, Arun, Avon* (Devon), *Avon* (Hants), *Calder, Camel, Cleddy, Clwyd* and *Elwy, Conway, Coquet, Dart, Dee, Derwent* (tributary of Trent), *Derwent* (Yorkshire), *Dovey, Dwyryd, Ellen* (Cumberland), *Erme* (Devon), *Exe, Fowey, Frome* (Dorset), *Glaslyn, Gwyfrai, Itchen, Mawddach, Nidd, Okement, Otter, Ouse* (Yorkshire), *Ribble, Sciont, Severn, Stour* (Dorset), *Stour* (Kent), *Swale, Tamar* and *Plym, Taw* and *Torridge, Tees, Teivy, Test, Towey, Trent, Tyne, Ure, Usk* and *Ebbw, Wear, Wharfe, Wye* and *Yealm* (Devon).

From the replies received by the Inspectors to questions Nos. 8 and 12 of a series they had addressed to the different Boards of Conservators, and from their own personal examination, it appears that of these fully

twenty-five rivers, all described in the Inspectors' previous
reports as variously polluted, still continued subject to the
same various forms of pollution—namely, from *lead
mines, sewage, tanneries, naptha, petroleum,* and *chemical
works, carpet manufactories, collieries, paper-mills,* and
from *vitriol, gas-tar,* &c., &c., such rivers being the
Calder, Camel, Cleddy, Dart, Dee, Derwent (tributary of
Trent), *Dovey,* with its tributary, the *Twymin, Exe,
Fowey, Ribble, Severn, Swale, Tamar* and *Plym, Tees,
Towey, Trent, Tyne, Wear, Usk* and *Ebbw., Wharfe,
Wye,* and *Yealm;* the Inspectors adding that the *Ouse*
(Yorkshire) was polluted near Linton Weir by water in
which *flax* had been *steeped,* the Devonshire *Avon* and
Erme by a new *sail-cloth factory,* and that at the outfall
of the main drainage an *offensive mud-bank* was accumu-
lating in the *Thames.*

As regards the *Calder,* Mr. Buckland, struck with its
frightful state of pollution, in reference to that river
says, at page 21 : " I went a long way up the River
" *Calder,* a fine tributary of the *Ribble.* The water here
" is wholly unfitted for fish to live in. It is a question
" not of fish alone, *but of the health of the inhabitants on
" the banks,* and the *population* on this river is *very
" dense.*"

Of the way in which the two lead mines on the
Twymin (the tributary of the *Dovey*)—viz., the Dylilfe
and Sir John Conroy's—pollute that stream and the
Dovey the Inspectors again speak in terms of strong con-
demnation, Mr. Buckland at page 14 saying that—

" There can be no doubt whatever but that these two
" mines are doing *an immense amount of injury to the
" fish* in the river, and I have written to the proprietors
" of them, calling their attention to the facts, and earnestly
" requesting them to take measures to keep the lead
" washings out of the river, and I do trust they will
" follow the good example shown by Mr. Beaumont in

" his lead mines on the *Wear* and *Tyne*. The very day
" before I was there almost every pool in the *Dovey*
" contained dead sewen, and a large number were brought
" to me."

And Mr. Walpole, at page 68, observes ;

" The *Dovey*, naturally one of the most important
" rivers in the kingdom, 35 miles long, and with a catch-
" ment basin of 264 square miles, is being terribly
" injured by two mines on one of its tributaries (the
" *Twymin*). When I was in Merionethshire last August
" the fish were lying dead in the river, *poisoned by pollu-*
" *tions, miles below the point at which the "hush" enters*
" *the river.*"

APPENDIX.

Statement of the efforts made between 1855 and 1868 to obtain from Parliament a general and effectual Law against the Pollution of Streams.

In 1855, when the "*Nuisances Removal Bill*" of Sir Benjamin Hall (the late Lord Llanover) was passing through Committee, Mr. Adderley, Mr. Henley, and Lord Robert Grosvenor attempted, but unsuccessfully, to get Clause 24 (which imposed a penalty of £200 for the discharge of gas refuse into the rivers) extended, so as to include* *all manufacturing refuse* of a foul, and poisonous nature.

1855.
Attempt to get Clause 24 of Nuisances Removal Bill extended.

Early in 1861 a number of noblemen and gentlemen associated themselves together under the name of the "Fisheries Preservation Association," for the purpose (among other objects they had in view) of obtaining an enactment against the "poisoning and polluting of rivers."

1861.
Formation of the Fisheries Preservation Association.

In the Session of this year (1861), through the instrumentality in a great measure of this Association, the Salmon Fishery Act for England and Wales (24 and 25 Vic., cap. 109) was obtained, and it contained a clause (Section 5) which at the time it was hoped would prove effectual in preventing the pollution of rivers.

1861.
Passing of Salmon Fishery Act.

Experience having shown, however, that the Act (a most valuable and effective one in other respects) was inoperative as regarded such pollutions, in August, 1863, a

1863.
Joint Deputation to Lord Palmerston.

* *Hansard*, 3rd Series, vol. 139, pp. 671-672.

joint deputation from the Sanitary Associations of Great Britain and the Fisheries Preservation Association went up to Lord Palmerston, and urged his Lordship to introduce a Government measure to put an effectual stop to the nuisance.

1864.
Joint Letter to Lord Palmerston.
On the 4th March, 1864, the same Associations, by their Presidents and Vice-Presidents, Lords Ebury and Shaftesbury, and Lords Saltoun and Llanover, addressed to Lord Palmerston a joint letter (from which extracts have been made at pp. 12—14 *ante*), " entreating his Lordship " to lose no time in proposing such measures as might seem " best adapted to prevent the spread of this enormous evil."

1864-5.
Memorials to Home Secretary.
In the same year (1864) and beginning of 1865, the boroughs of *Nottingham, Sheffield, Birmingham, Manchester, Preston, Coventry, Derby, Wolverhampton, Bath, Huddersfield, York, Stockport, Cheltenham*, and *Oxford*, and the *Rotherham* and *Kimberworth* Board of Health, all memorialised (as set forth at pp. 16-20) the Home Secretary to carry out the recommendation (before given at p. 16) of the Committee of the House of Commons of 1864, viz., " that the important object of freeing the entire basins of " rivers from pollution should be rendered possible by " general legislative enactment."

1865.
Petitions to Parliament.
In this year (1865) also petitions against the pollution of rivers, most numerously signed, were presented to Parliament by the late Lord Llanover, then President of the Fisheries Preservation Association, in the one House, and by Mr. Martin Tucker Smith in the other, from *Shrewsbury, Richmond, Twickenham, Machynlleth,* &c., &c.

24 Feb. 1865.
Earl of Longford, in the Lords, inquires if Government intends to
On the 24th February, 1865, the Earl of Longford, in the House of Lords, called the attention of Government to the beforementioned recommendation of the Commons' Select Committee of 1864, and enquired whether it was the intention of Government to carry into effect that re-

commendation, the reply of Lord Granville, on the part of the Government, being that "the Home Office was in "communication with Lord Robert Montagu, who had "introduced a measure on the subject into the other "House, and he (Lord Granville) hoped a satisfactory "measure would be framed." *carry out recommendations of Committee of 1864. Hansard, 3rd Series, vol. 177, pp. 636 to 641.*

On the 8th March following, Lord Robert Montagu, in a most able and exhaustive speech, moved the second reading of the "River Waters Protection" Bill.

This Bill, however, owing to strong objections being felt by the House to its machinery, was, in deference to a generally expressed wish, withdrawn, though at the same time the enormous extent and dangerous nature of the nuisance it sought to arrest, and the pressing necessity for legislative action, were admitted to have been fully established by Lord Robert Montagu's masterly exposition. *March, 1865. Lord Robert Montagu moves second reading of River Waters Protection Bill. Hansard, 3rd Series, vol. 177, p. 1309.*

On the 17th of the same month a deputation from the Fisheries Preservation Association, including in it Lords Robert Montagu and Ebury and other influential persons, had an interview with the Home Secretary, when it most earnestly represented to Sir George Grey the urgent need there was for the Government, on the supreme ground of the public health, and in the interest of the fisheries, taking prompt and energetic steps to abate an evil which had now become so vast and in every way so baneful. *17th March, 1865. Deputation from Fisheries Preservation Association to Home Secretary.*

The result of this interview was the issuing by Sir George Grey on the 18th May following of a Royal Commission "to inquire into the best means of pre-"venting the pollution of rivers. *18th May, 1865. Issue of the Royal Commission on the Pollution of Rivers.*

Extracts from the several reports of this Commission on the *Thames*, the *Lea*, and the *Aire* and *Calder* have been given at pages 25—27, 28—30, 32—38.

In the Session of 1866 the "Thames Navigation Act" (29 and 30 Vic., cap. 89) passed, containing certain *1866. 29 & 30 Vic., c. 89. Thames Navigation Act (short title).*

D

provisions for freeing the river between Staines and Cricklade, in Wilts, near its source, from sewage and noxious and offensive refuse; and in the same Session also another Act (29 and 30 Vic., cap. 319) was passed, intituled "An Act for the purification of the River " Thames, by the diversion therefrom of the sewage of " *Oxford, Abingdon, Reading, Kingston, Richmond,* " *Twickenham, Isleworth* and *Brentford,* and for the col-" lection and utilization of the sewage."

1866.
29 & 30 Vic.,
c. 319.
Thames Puri-
fication Act
(short title).

This latter Act empowered certain persons therein named, who were willing at their own expense to divert the sewage from the river, to incorporate themselves into a company for that purpose. No company, however, appears to have been formed nor anything done under the powers of this Act.

1867.
30 Vic., c.
101.
" Thames
Conservancy
Act" (short
title).

In 1867 an Act was passed conferring on the Conservators of the Thames the same powers for preventing the pollution of the river between *Staines* and the *western boundary* of *the metropolis,* as had been conferred on them by the Thames Navigation Act of the previous year, as regarded the portion of the river between *Staines* and *Cricklade* near its source; but it would seem from the subsequent statements made by Sir George Bowyer (on the motion of Mr. Cave for leave to bring in a Bill for the Conservancy of the River Lea—(*Times* Report, 21st February, 1868) that this Act and the previous Act of 1866 were quite inoperative as regarded the exclusion of sewage from the river, the towns on it, Sir George remarked, "having done nothing to exclude it, and " they declared that they could not be compelled to exclude it."

7th June,
1867.
Salmon Fish-
ery Congress
at South
Kensington

On the 7th June of this year the great Salmon Fishery Congress assembled at the Horticultural Gardens, South Kensington. It was presided over by Earl Percy (now the Duke of Northumberland), and consisted of noblemen and gentlemen from all parts of the kingdom, interested in the fisheries as conservators, proprietors, &c. Mr. Frank Buckland and Mr. Spencer Walpole, the Inspectors of Salmon Fisheries for England and Wales, also attended it, as did Major Scott and Captain Spratt, the Special Commissioners of Fisheries, the President (Lord de Blaquiere), and various members of the Council of the Fisheries' Preservation Association being likewise present and taking active part in the proceedings.

The subject of the pollution of rivers and its destructive effect on the public health and on the fish, and the inefficacy of existing laws to meet the grievance having been very largely dwelt upon by the Chairman, (Lord Percy), and by every speaker who took part in the discussion, the following resolution was moved by Mr. Higford Burr (one of the Council of the Fisheries Preservation Association) and carried unanimously :—

" That as various rivers are seriously injured both by " liquid and solid poisonous matters, and as it is neces- " sary for the public health and supply of food for the " people, that the pollutions should be prevented, further " legislation is urgently needed."

Unanimous Resolution passed that further legislation is urgently needed to prevent pollutions.

On the 6th of the following August another deputation from the Fisheries Preservation Association, including Lord Northwick and other influential persons, members of Parliament and others, waited on Mr. Gathorne Hardy, the Secretary of State, and again earnestly pressed on the attention of Government, the absolute necessity that existed of putting a stop to the pollution of the rivers by some effectual measure to be introduced by the Government. Whereupon the deputation had the satisfaction of receiving from the Home Secretary the assurance " that he did not intend to continue the investi- " gation, as he believed the experience gained by the " inquiries into a few rivers would govern the whole,

6th August, 1867. Deputation from Fisheries Preservation Association to the Home Secretary.

Declaration of Home Secretary, &c. [The *Field* Report, 10th Aug., 1867.

" and that he would, during the forthcoming recess, give
" the whole subject his best consideration."

1868.
Lea Conservancy Bill.
Times Report, 21st
Feb., 1868.

On the 20th February of the present session (1868) the
Right Hon. Stephen Cave, the Vice-President of the
Board of Trade, on the part of the Government, brought
forward the River Lea Conservancy Bill, based on the
recommendations respectively of the Committee of 1866
on the East London Water Bills and of the Pollution of
Rivers Commissioners, the object being the preservation
of the purity of the water of the Lea. The bill was read
a first time, and Mr. Cave was to give notice of the second
reading.

Ibid.

On this occasion Mr. Powell said " he hoped that
" would not be the only bill of its class the Government
" would introduce during the present session, but that
" they would grapple with the case of the *Aire* and *Calder*,
" which urgently demanded attention, the population in the
" basin of which exceeded a million, whereas that occu-
" pying the basin of the Lea was only about a quarter of
" a million."

Times Report, 25th
Feb., 1868.

On the 24th of the same month Mr. Candlish, the
member for Sunderland, inquired in the House of Commons of the Home Secretary if he meant to introduce a
measure this session to prevent the pollution of rivers, or
to prevent solids being deposited in rivers.

The reply of Mr. Gathorne Hardy was—

" That he was not prepared to legislate on the subject
" this session.

" It was his intention to appoint a new Commission to
" inquire into those parts of the subject which had not
" yet been investigated, and he did not think it advisable
" to deal with only a part of the question."

FISHERIES PRESERVATION ASSOCIATION.

REPORT FOR THE YEAR ENDING MAY 1st, 1868.

THE Council beg to submit to the Members their Report for the past year.

LORD DE BLAQUIERE having expressed a wish to retire from the Presidency of the Association in favour of HIS GRACE THE DUKE OF NORTHUMBERLAND (who, as Earl Percy, had already last year been elected a Member of the Council), and the DUKE having accepted the post, the Council have much pleasure in stating that HIS GRACE has been chosen President accordingly,

LORD DE BLAQUIERE at the same time, the Council are extremely happy to say, consents to give to the Association, as Vice-President, the benefit of his continued services and influence.

It is with much gratification the Council announce the accession to their Board of LORD ABINGER, who was unanimously elected a Member of the Council in February last.

They also have to announce that, in appreciation of their valuable services to important Salmon Fisheries, JOSEPH DODDS, Esq., Honorary Secretary of the Tees Salmon Fishery Landowners' Association, and John Lloyd, Esq., of Huntington Court, Hereford, a Conservator of the Wye, Usk, and Ebbw Fisheries, have been elected Honorary Members of the Association.

With respect to the great object which the Council have so constantly held in view, and so long striven to effect—the prevention of the pollution of the Rivers of the kingdom—they grieve to say that they see no chance *this Session* of the hope which they ventured to express in their last Report being fulfilled, viz., of obtaining a remedial measure for that evil.

The state of the public business in Parliament, the position of Parliament itself, and the unexpected inaction of the Government as represented by the Home Secretary, alike preclude the expectation that any such measure can possibly be introduced during the present Session.

Steadfast, however, to their purpose of inducing, if possible, the Government to bring in a Bill on the subject, on the 6th August last a Deputation from the Association, introduced by their President (LORD DE BLAQUIERE), and which had the advantage of being joined by LORD NORTHWICK, MR. CANDLISH, COL. SYKES, and other Members of Parliament and persons of influence, had an audience of the HOME SECRETARY, when the question, both on the paramount ground of the public health and as it affected the Fisheries, was most earnestly pressed on the consideration of MR. GATHORNE HARDY, and the reply of MR. HARDY to the Deputation was such as to create the belief that the Government would not let the present Session pass without taking legislative action in the matter, that reply being that " *he did not intend to con-* " *tinue the investigations, as he believed the experience gained by the inquiries into a few rivers would* " *govern the whole.*"

A few days after this Deputation the Commissioners on the Pollution of Rivers made their Report to the Home Secretary on the state of the *Aire* and *Calder.* Of this Report it may suffice to say that, dreadful and disgusting as was the picture presented in the Commissioners' former Reports of the pollution of the *Upper Thames* and the *Lea*, the description they give of the polluted condition of the *Aire* and *Calder* far surpasses that picture in its worst and most revolting features.

Notwithstanding that Report, however, and the auspicious answer to the Deputation just stated, to the Council's great surprise and mortification, the HOME SECRETARY, when questioned, on the 24th February last, on the subject by the Member for Sunderland (MR. CANDLISH), " answered that " *he was not prepared to legislate in the matter this Session, and that he was* " *about to appoint a fresh Commission to continue the inquiries;*" and, since that declaration, the Right Honourable Gentleman has appointed such fresh Commission.

2

The Council will not relax their efforts in the cause, but it seems to them that the best hope of obtaining a measure adequate to meet this terrible evil of pollution of streams rests now on the public voice making itself promptly, powerfully, and decisively heard by the Government.

To that end, and in order to make the fearful state of our rivers more widely known, the Council have prepared and are now disseminating a pamphlet, containing in a concise form all the necessary facts relating to the question, extracted from the voluminous Blue Books, &c., of the last thirteen years.

Although the Council have not yet succeeded in accomplishing the main object of their endeavours, viz., the obtaining of an Act to prevent the Pollution of Rivers, they are, nevertheless happy to say that the past year has by no means been barren of results advantageous to the Fisheries at large, or promising to be so, as the following proceedings will show :—

In the first place, on the 7th June of last year, a great Salmon Fishery Congress assembled, under the presidency of Earl Percy, at South Kensington. This Congress included various noblemen and gentlemen, great Fishery proprietors, Conservators, and others from all parts of the kingdom. It was attended by Mr. Frank Buckland and Mr. Spencer Walpole (the Government Inspectors of Salmon Fisheries for England and Wales), by the Special Commissioners of Fisheries, and by the President (Lord de Blaquiere), and several members of the Council of the Fisheries Preservation Association, all of whom took an active share in the proceedings.

A great portion of these proceedings having been devoted to the consideration of the pollution of rivers,—Mr. Bigford Burr (of the Council of this Association moved and carried unanimously a resolution in the following terms:—

" That as various rivers are seriously injured by liquid and solid poisonous matters, and
" as it is necessary for the public health and the supply of food for the people, that the
" pollutions should be prevented, further legislation is urgently needed."

From the discussions that arose and the interchange of suggestions which took place, not merely as regarded the evil of pollutions but on all points affecting the interests of the Fisheries, at a Congress so influential and possessing such special and practical knowledge of the subjects brought under deliberation, it is impossible to doubt that great benefit must ensue to the Fisheries.

In the next place, through the great exertions mainly of Lords Abinger and Wynford, aided by their lordships' friends in the Upper House, and assisted by Lord de Blaquiere and his friends, a very pernicious measure of Lord Cranworth relating to the Irish Fisheries was in July of last Session thrown out in the House of Lords.

The importance of its rejection may be estimated by the fact that had that Bill passed into law it would have had the effect of undoing much of the beneficial legislation of late years by re-establishing in Ireland those destructive engines which the friends of the Fisheries had been at such great pains to get abolished.

Lastly, on the 2nd April of the present year a deputation of a very influential character, consisting of Lords Abinger, Airlie, Seafield, Colville, and Saltoun, and other large Scotch proprietors, &c. (and of which various members of the Council of this Association made a part), having been formed by Lord Abinger, was introduced by his lordship to the Home Secretary for the purpose of ascertaining from the Right Honorable Gentleman what steps Her Majesty's Government proposed taking this Session for the amendment of the Scotch Fisheries Act.

Lord Abinger having explained that the amended Act desired was one based on the Scotch Bill of 1866, which had passed the House of Lords and that the chief amendments required were :—

1st. The appointment of permanent Inspectors empowered to carry out such alterations in the bye-laws as might from time to time be found expedient, and whose duty it should be to report annually the general condition of the Scotch Salmon Fisheries.

2nd. The confirmation of the bye-laws, already passed by the Commissioners in order to set at rest doubts arising from the peculiar wording of the Act of 1862 as to the legality of those laws.

Mr. Gathorne Hardy in reply informed the Deputation that "the whole subject of the " amendment of the Scotch Salmon Fishery Acts was then under the consideration of the " Government, and was engaging the attention of the Lord Advocate."

The Council cannot close their Report without congratulating the members on the fact that the two last valuable Reports (July, 1867, and March, 1868) of the Inspectors, declare that the English and Welsh Salmon Fisheries are on the increase, four and twenty rivers being specially named in the first of those Reports as exhibiting " great and marked improvement."

Among other kind contributions to the funds of the Association since the last Report the Council have the pleasure of acknowledging the following donations—namely :—Earl de Grey and Ripon, £10 10s.; Lord Northwick, £10 10s.; Lord Ebury, £10 10s.; J. H. Arkwright, Esq., £10; Archibald Cockburn, Esq., £10 10s.

Annexed is the Honorary Treasurer's Account of Receipts and Payments for the past year, shewing a balance to the credit of the Association on the 1st May inst. of £183 1s. 9d.; but this balance will be much diminished by the stationery, printing, and some other expenses of the past year, while those of the current year will be more than usually large.

The Council, in concluding their Report, and in reminding the Subscribers that their Subscriptions became due on the 1st of the present month, would earnestly impress upon every Member the strong necessity of his kindly continuing his active support to the Association.

DE BLAQUIERE,
President.

Fisheries Preservation Association,
25, Lower Seymour Street,
Portman Square, W.
1st May, 1868,

STATEMENT OF RECEIPTS AND EXPENDITURE,

For the Year ending 1st May, 1868.

RECEIPTS.		£	s.	d.
1867—May 1.				
To Balance to the credit of the Association as per last Report		116	0	7
To Donations and Subscriptions received since and up to 1st May, 1868		119	3	0
		£265	**3**	**7**

EXPENDITURE.		£	s.	d.
1867—June 22.				
Paid Mr. Smith (Stationer's Account) for Printing, Stationery, Postage Stamps, &c., for 1866		48	11	10
Paid Assistant Secretary (Mr. Kinloch) 13 weeks' and one day's Salary, to 6th August, 1867, at £1 10s.	£19 15 0			
Do. do. 6 weeks and 4 days' do. to 29th March, 1868	10 0 0	29	15	0
Paid Collector's Commission on Town Subscriptions		0	19	0
Petty and Miscellaneous Expenses...		2	13	0
1st May, 1868.				
To Balance to credit of the Association		183	1	0
		£265	**3**	**7**

DONORS, SUBSCRIBERS, AND HONORARY MEMBERS, PAST AND PRESENT.

The names of such as are deceased being in *Italics*.

Abinger, Lord, 48, Chester Square, S.W.
Airey, B. Esq. Lloyds, E.C.
Arden, J. Esq. Rickmansworth-park, Herts.
Arkwright, J. H. Esq. Hampton Court, Leominster.
Baillie, Capt. Duncan, 23 Queen's-gate-terrace, W.
Barlow, E. Esq. Bolton-le-Moors.
Barnes, J. Esq. Charlewood, Rickmansworth.
Barry, Herbert, Esq. 67 Belsize-park, Hampstead.
Baxendale, J. Esq. Gresham-street, E.C.
Beaumont, Wentworth, Esq. M.P. 144 Piccadilly.
Benthall, F. Esq. 23 Old-square.
Beresford, D. Packe, Esq. M.P. 32 Devonshire-place, W.
Bernard, J. Esq. Church-place, Piccadilly.
Berwick, Lord, Attingham Hall, Shrewsbury.
Berwick Shipping Company, Berwick-on-Tweed.
Bicknell, Henry, Esq. 28 Upper Bedford-place, W.C.
Blenkinsop, J. Esq. North End House, Watford.
Boyden, T. W. Esq. 2 Woodville-terrace, Mildmay-park.
Brandreth, Chas. Esq. Army and Navy Club.
Brandreth, E. Esq. 3 Eaton-square.
Breadalbane, Earl of, Taymouth Castle, Perthshire.
Briggs, H. R. Esq. Waldrons, Croydon.
Bryant, Dr. 23a Sussex-square, W.
Buckland, F. T. Esq. Athenæum Club.
Bulteel, John, Esq. Bantteet, Ivy-bridge, Devon.
Burr, Higford, Esq. 23 Eaton-place, S.W.
Bute, Marquis of, 83 Eccleston-sq., Cardiff Castle, Cardiff.
Cannons, P. T. Esq. 1, Vanbrugh-park, Blackheath
Carpenter, Capt. R A.
Carruthers, Richard, Esq. Eden-grove, Carlisle.
Chadwick, F. Esq. The Hermitage, Preston, Lancashire.
Chester, C. Esq. Oriental Club.
Chetty, Edw. Esq. Temple.
Chitty, Thos. Esq. Temple.
Clutterbuck, T. Esq. Micklefield Hall, Watford.
Cocks, T. S. Esq. 43 Charing-cross.
Cockburn, Archibald, Esq. 60 Mark-lane, E.C.
Cockerell, S. P. Esq. 45 Hertford-street, W.
Collier, W. J. Esq. Woodtown, Horrabridge, Devon.
Colville, Lord, 42 Eaton-place.
Crawshay, E. Esq. Otheroe House, Market Harborough.
Crockford, John, Esq (Editor of the *Field.*)
Crowe, Capt. M. Newby-bridge, Newton-in-Cartmel.
Cunninghame, Jas. Esq. Prince's-park, Liverpool.
Dacre, Lord, 40 Belgrave-square.
Daniell, Edw. Esq. Thropham Hall, Norfolk.
Darke, —, Esq. (per J. Hobbs, Esq.)
De Blaquiere, Lord, 9, Stratford-place, W.
De Grey and Ripon, Earl, 1 Carlton gardens.
Denison, A. Esq. 6 Albemarle-street.
Derwent Fly Fishing Club, Sheffield.
Dodds, Josh. Esq. (Hon. Sec. Tees Salmon Landowners' Association), Stockton-on-Tees, Honorary Member.
Donoughmore, 4th Earl of, 52 South Audley-street.
Douglas, Col. John, Glenlinart, Argyleshire.
Dovey Club, Machynlleth.
Draper, Jno. Reform Club.
Ducie, Earl of, 1, Belgrave-square.
Duff, B. Abercromby, Esq. M P. 40 Mount-street, Grosvenor-square.
Dynevor, Lord, 19 Prince's-gardens, W.
Ebury, Lord, 107 Park-street, W.

Edwards, Lt. Col. M.P. 32 Dover-street, W.
Ellesmere, Lord, Esq.
Farquhar, Sir Minto, Bart. 4 Berkeley-street, Piccadilly.
Fawcett, Henry, Esq. M P. United University Club.
Fielden, Sir W. H. Feniscowles, Blackburn.
Fife, Earl of, Cavendish-square.
Fisher, Jno. Esq. Reform Club.
Fisher, W. W. Esq. 16 Dorset-square, N.W.
Foott, Edward, Esq. Gortmore, Kanturk.
Francis, Francis, Esq. The Firs, Twickenham.
Fraser, Col. the Hon. Royal Horse Artillery.
Fraser, Major Tytler, of Aldowrie, Inverness.
Freeman, J. Esq. Harrow.
Freeman, T. A. Esq. Christchurch, Oxford.
Freeman, W. Esq. 27 Millbank-street, S.W.
Frere, C. Esq. Western Life Office, Parliament-street.
Fuller, P. H. Esq. Stoke Pogis, Bucks.
Garnett, W. J. Esq. M.P.
Geldart, T. C. Esq. Trinity Hall House, Cambridge.
Goddard, A. C. Esq. Casley-green, Rickmansworth.
Goodhart, Jos. H. Esq. Manor House, Tooting.
Gore, Sir St. George, Bart. 15 South Audley-street.
Grant, Col. 3 St. George's-place, S.W.
Grant, Sir G. Macpherson, Bart. Ballindalloch Castle, Elgin, N.B.
Greenwood, Geo. Esq. Alramant, Builth.
Gregson, S. Esq M.P. 32 Upper Harley-street.
Griffith, Davies H. Esq. Taerhün, Conway.
Guest, Merthyr, Esq. Methyr Tydvil
Harrison, J. Esq. 43 Leeterrace, Blackheath.
Haswell, Millborough, Esq. Ludlow.
Hawkins, Capt. Frank, I.N. Army and Navy Club.
Hawley, Sir J. Bart 34 Eaton-place.
Hindley, W. Esq. Grenaly House, Croydon.
Hobbs, J. Esq. G. P. O. St. Martin's-le-Grand, E.C.
Holyoake, Henry, Esq. Travellers' Club.
Holyoake, Capt. ditto.
Hood, Peter, Esq. M.D. 23 Lower Seymour-street, W
Hopkinson, J. Esq. 255 Regent-street, W.
Hopwood, J. T. Esq. M.P. Carlton Club.
Howard, H. R. Esq. Watford.
Hunt, J. Esq. 107 Victoria-street, S.W.
James, Sir Henry, Ordnance Survey Office, Southampton.
Jackson, Capt. late 16th Lancers, Army and Navy Club.
Jackson, Major, Warwick Hall, Carlisle.
Keane, Capt. The Hon G. D. R.N.
Keane, Col. The Hon. H. F.—R.E. 76 Jermyn-street.
Keane, the Hon. J. Arbuthnot, Castletown House, Wexford.
Kent Angling Association, Kendal.
King, Jonathan, Esq. Wiggan Hall, Watford.
Kinrosshire Fishing Club, Kinross, N.B.
Knight, J. W. Esq. 110 Great Portland street
Lake, Capt. Percy, Preston, Lancashire.
Lanure, Lord, 9 Great Stanhope-street, W.
Lee, John, Esq.
Lee, Capt. Vaughan, Army and Navy Club.
Liddell, Hon. H. M. 9 Mansfield-street, W.
Lindley, S. H. Esq. 19 Catherine-street, Strand.
Littledale, Geo. Esq Army and Navy Club.
Lloyd, John, Esq. Huntington Court, Hereford, Honorary Member.
Longford, Earl of, 24 Bruton-street.

Lucas, Edmund, Esq. Milbank-street.
Macnaghten, Sir E. W. Bart. 18 Eaton-square.
Malmesbury, Earl of, 19 Stratford-place, W.
Marriott, Col. Fielding, R.A.
Martin, Major, Upton-on-Severn.
Maunsell, Capt. Cockayne, Thorpe Malsor. Kettering.
Maunsell, Capt. Cullen, ditto.
Mount Charles, Earl of, 60 Rutland-gate, W.
Miller, W. Esq. jun. 4 St. Helen's-place, E.C.
Neale, Vansittart, Esq. Western Lit. Office, 7 Waterloo-pl, W.
Nicholay, J. A. Esq. jun. Cumberland-mills. Isle of Dogs.
Norman, Henry, Esq. 11 Henrietta-street, Cavendish-sq.
Northumberland, Duke of, Northumberland House, S.W.
Northwick, Lord, 22 Park-street, W.
Oawry, F. Esq. 12 Queen Anne-street, W.
Owen, A. Mostyn, Esq. Erwy, Rualson.
Palmer, J. Carrington, Esq. (per Major-General Ramsay).
Parry, Richard, Esq. Army and Navy Club.
Peach, Major, Army and Navy Club.
Pennell, C. H. Esq. Weybridge.
Perceval, P. Esq. 15 South Audley-street.
Perkins, Algernon, Esq. 81 Harley-street.
Phelps, J. L. Esq. Plassy, Limerick.
Phillips, J. H. Esq. Nawton, Yorkshire.
Pipon, Col. Deerwood, Crawley, Sussex.
Bedcarth, Lord, Merton House, Berwick-on-Tweed.
Recombe, Lord, Royal Yacht Club, Cowes.
Porter, E. Esq. Sheffield.
Powerscourt, Viscount, 37 Grosvenor-square.
Pretyman, Col. Army and Navy Club.
Price, Bonamy, Esq. 11 Prince's-terrace, S.W.
Prittie, Hon. Francis, The Lodge, Clonmel.
Pryse, Col M.P. Army and Navy Club.
Ramsay, Major-Gen. 1 Sussex-square, W.
Reeve, John, Esq. Albemarle-street.
Richardson, Wm. Esq. Hill House, Hatfield.
Rickmansworth Fishing Club.
Bishby, Gen. Wm. Malta.
Robinson, F. J. Esq.
Robinson, C. Esq. 65 Basinghall-street. E.C.
Rooper, Geo. Esq. Nascott House, Watford.
Rutter, Isaac, Esq. Glebe-lands, Mitcham.
Ryedale Fishing Club, Yorkshire.
Saltoun, Lord, Ness Castle, Inverness.
Schroeder, H. S. Esq. 16 Carlton-road, Maida-vale.
Scott, Jas. Esq. Roding House, Woodford, Essex.
Seager, J. L. Esq. South Lambeth.

Sharpe, Edw. Esq. Llanwrst, N. Wales.
Shaw, Capt. N. Armathwick, Carlisle.
Shropshire Association of the Severn, Shrewsbury.
Sibthorp, Major, Canwick Hall, Lincoln.
Sinclair, W. Esq. Drumbeg, Donegal.
Skrine, H. D. Esq. Warleigh Manor, Bath.
Smith, Arthur, Esq. 24 Wilton-street, S.W.
Smith, C. J. Esq. Park House, Isleworth.
Smith, Martin T. Esq. 13 Upper Belgrave-street.
Smith, M. R Esq. ditto.
Smith, Oswald A. Esq. 1 Lombard-street, E.C.
Smith, Dr. Tyler, 21 Upper Grosvenor-street, W.
Sparrow, C. Esq. 24 Pembridge-crescent, W.
Stewart, H. G. Murray, Esq. Killebegs. Donegal.
Sylvester, Edw. Esq. North Hall, Chorley.
Taylor, Sir C. Bart. 20 King-street, St. James's.
Templetown, 2nd Viscount, Albany, Piccadilly.
Templetown, 3rd Viscount, Devonport.
Tennant, J. R. Esq. The Hall, Kildwick, Leeds.
Thornhill, W. P. Esq. M.P. 44 Eaton-square.
Thoroton, Capt. J. Army and Navy Club.
Thruston, C Esq. Talgarth Hall, Machynlleth.
Traherne, Major.
Tredegar, Lord, 39 Portman-square.
True Waltonian Society, "The Crown," Pentonville-hill.
Vansittart, Gen. Esq. 34 Bryanston-square.
Vansittart, Spencer, Esq. Castle Connell, Limerick
Vaughan, Edward, Esq. Rheola, Neath.
Vaughan, Nash Edward, Esq. ditto.
Warwick, Benj. Esq. Englefield-green.
Wemyss, Earl of, Gosford House, Haddington.
Wharncliffe, Lord, Wortley Hall, Sheffield.
Whimper, Col. 97 Ebury-street, S.W.
Whitaker, T. S. Esq. Everthorpe Hall, East Yorkshire.
White, Col. Sligo.
White, John Bazley, Esq. Blackheath.
Whittingstall, Capt. 35 St. George's-road, S.W.
Wickham, H. W. Esq. M.P. 3 Chapel-street, W.
Willaume, Tanqueray, Esq. 24 Chester-terrace, Regent's-pk
Willis, Henry, Esq. Old Windsor, Berks.
Wilson, G. P. Esq. 3 Chepstow-terrace, W.
Wotton, H. R. Esq. 17 Cavendish-square.
Wright, R. Esq. Harlesden-green.
Wynne, Brownlow, Esq. Garsthewin, Abergele.
Youl, J. A. Esq. Waratah House, Clapham-park.
Young, R. Esq. M.P. 19 Eccleston-street. S.W.
Zetland, Earl of, Upleatham, Yorkshire.

FISHERIES

Preservation Association.

REPORT

For the Year ending May, 1868.